The Natural Beauty & Bath Book

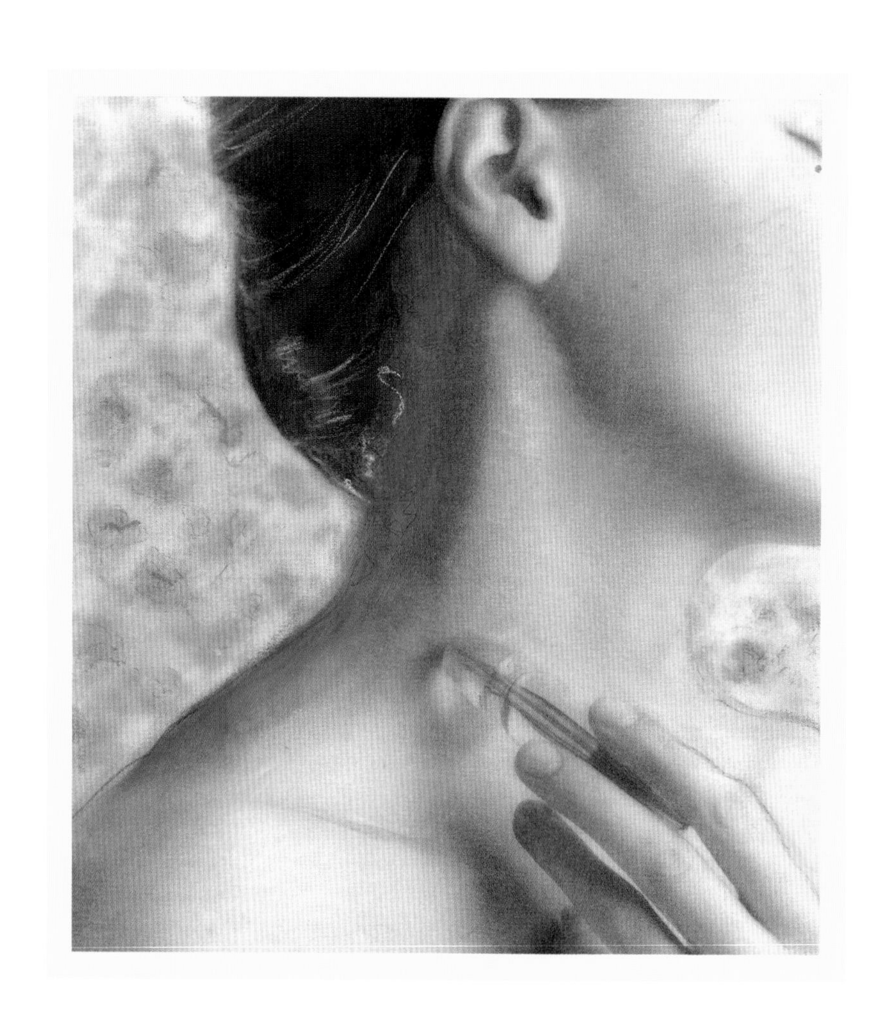

The Natural Beauty & Bath Book

Nature's Luxurious Recipes for Body & Skin Care

Casey Kellar

Lark Books
A division of Sterling
Publishing Co., Inc.
New York

Editor: Leslie Dierks
Art Director: Celia Naranjo
Photography: Evan Bracken
Color creations on chapter opening pages: Brigid Burns
Production: Celia Naranjo

Library of Congress Cataloging in Publication Data
Kellar, Casey.
The natural beauty & bath book : nature's luxurious recipes for body & skin care / Casey Kellar. —1st ed.
p. cm.
Includes index.
ISBN 1-57990-178-6
1. Cosmetics. 2. Toilet preparations. I. Title.
TP983.K425 1997 97-16341
668'.5—dc21 CIP

10 9 8 7 6 5 4 3

Published by Lark Books, a division of
Sterling Publishing Co., Inc.
387 Park Avenue South, New York, N.Y. 10016

Distributed in Canada by Sterling Publishing,
c/o Canadian Manda Group, One Atlantic Ave., Suite 105
Toronto, Ontario, Canada M6K 3E7

Distributed in Australia by Capricorn Link (Australia) Pty Ltd., P.O. Box 6651, Baulkham Hills, Business Centre NSW 2153, Australia

If you have questions or comments about this book, please contact:
Lark Books
50 College St.
Asheville, NC 28801
(828) 253-0467

Printed in China by Oceanic Graphic Printing Productions Ltd.

ISBN 1-57990-178-6

Dedication

This book is dedicated to my family: To my grandfather, G.D. Field, who spurred my early interest in natural formulas and who left me all of his wonderful old books and notes. To my father, Donald, for his strength and tenacity, and to my mother, Irene, for her support. To my children, Justin and Tiffann, for allowing me to experiment on them and for helping make our family business a success. To my beautiful granddaughter, Kiley, who keeps alive the wonder and joy in life and reminds me to "keep it simple." And most of all, to my wonderful husband, Byron, for his constant encouragement and wholehearted help in every step.

Acknowledgments

Many thanks to everyone who assisted us with the photography for this book:
Janet Frye at The Enchanted Florist
Ned Gibson at B.B. Barns
Robert Glasgow at The Beaufort House
Tex Van Hoefen Harrison at Complements to the Chef
Ronnie Myers at Magnolia Beauregard's Antiques
Sue Wheeler at Mountain Food Products
Elena Wrightson at Body Care
all of Asheville, North Carolina.

Special thanks also to our models,
Phyllis Barnard, Evans Carter, and Marque Gritta.

Contents

Chapter One

Introduction

THROUGHOUT MY CHILDHOOD, my grandfather fascinated me with stories about the healing properties of herbs and natural folk medicine. A naturopath and chiropractor, he had a whole library of old alchemy books that were filled with descriptions of herbs and formulas for natural remedies. Today those books are part of my own library. Although some of the recipes seem antiquated and simple in view of how we now achieve emulsification and other effects, many formulas are still current. The chemistry of making basic soap, the use of specific herbs, and "scent therapy" haven't changed over the years, and some earlier theories are being rediscovered today.

Despite this background in natural therapies, I followed the path of conventional medicine, first as a phlebotomist, then as a radiology technician. While I was studying pharmacology, I began noticing that my daughter was allergic to *everything*, and I was having bouts of eczema, so I went back to grandpa's books to find some relief. After sorting out the myth from the fact, I found some very interesting answers. The old-fashioned formulas avoided harsh chemicals in favor of simple, natural ingredients, and—more important—the natural formulas worked!

In the following pages, you will find everything you need to know to make delightful and personalized bath, body, and other beauty products in your own kitchen. For many people, these homemade beauty formulas are preferable to the store-bought variety because all of the ingredients are natural substances that are known to be safe to the skin. Not one of the formulas in this book contains artificial preservatives or pH stabilizers, and for this reason, many of the products should be stored in the refrigerator. Some (such as the spa-style body wrap and some of the facial masks) are better made just for yourself and used right away; others (creams, lotions, salves, and after-shaves) have a moderate shelf life and can be given as gifts; still others (massage oils, lip balms, and soaps) are stable enough that they can be made up ahead of time and presented to family and friends on holidays and birthdays. Each recipe includes the shelf life and storage requirements of the finished product.

Today's hustle-bustle world is filled with stressful situations, and anything that encourages us to take the time to pamper ourselves is a welcome treat. These recipes will help ease you into a slower-paced, gentler way of life that is reminiscent of the days when kitchen alchemy was commonly used to enhance our beauty and the quality of our lives. As you make your own natural beauty products, you will experience the joy of discovery, and you will delight in the pleasure they bring as you use them yourself and share them with others.

History of Aromatherapy, Perfumes & Personal Care

Long before history recorded their activities, our ancestors discovered that burning certain aromatic plants and resins and inhaling their aromas had dramatic effects on their health and mental state. The fragrances of some plants caused drowsiness, and others created feelings of euphoria; some even warded off disease. These early forms of incense were used by tribal elders and other leaders in a variety of religious rituals and healing ceremonies.

Similarly, just as burning incense wasn't invented in the 20th century, neither was a concern for cleanliness. Personal hygiene has been important in many cultures throughout history. Perfumes have been found in the tombs of Egyptian pharaohs, who lived more than 3,000 years ago, and according to the recipes found among their inscriptions, the Egyptians used a wide variety of plant essences for healing, massage, and cosmetics. It is said that Cleopatra bathed in milk and honey to soften her skin and had her bath waters filled with fresh flower petals to give her skin a soft fragrance.

Many of the oils and spices used by the ancient Egyptians were imported from China and India, where their benefits were already well established. In China, incense was burned to soothe both mother and infant during childbirth, and some plant essences were believed to hold the key to immortality. India had an active market for perfumes and incense, and the early religious temples were constructed of sandalwood, which endowed them with a delicate fragrance.

During the Greek and Roman eras, herbs were studied extensively and their uses for medicinal purposes carefully documented. Hippocrates, the father of medicine, taught that the key to good health was a fragrant bath and massage each day. In Greek and Roman bathhouses, floating spices and fragrant flowers were placed in the bath, and scented oils were used to anoint the body and hair.

The Romans' knowledge of aromatic oils and herbal remedies was carried with them as they conquered much of Europe. After the fall of the Roman Empire and during the medieval period, however, it was left to Christian monks and individual village healers to maintain the herbal folklore that was passed from generation to generation. Soaps and bathing were a rarity, but scented oils were applied to the body, and aromatic sticks of wood were burned in the home to ward off disease. After the Crusaders began returning from the East with perfumes and exotic essences, these substances gained popularity among the wealthy, and France became the European center for perfumery.

In Europe and America, the Victorian era saw a renewed interest in a multitude of bath and beauty preparations. Rosewater and other floral waters became popular for keeping the skin toned and sweetly scented, and young ladies regularly dipped their fingertips into the fragrant waters, just in case their hands might be kissed. Light floral waters replaced regular water in the pitchers kept for daily washing. Herbs were sewn into pillows to inspire sweet dreams, and simple cosmetics, such as facial powder and rouge for the lips and cheeks, became widely available.

Today the average consumer has countless beauty preparations from which to choose. Unfortunately, many of these products contain strange ingredients with industrial-sounding names. A simple, wholesome alternative that is handmade with familiar, natural ingredients is indeed a gift, both for the person making it and the one using it.

Chapter Two

Ingredients, Supplies & Equipment

THE TOOLS AND EQUIPMENT needed for making your own natural beauty products are very simple and probably already reside in your kitchen. Don't let yourself be intimidated by the number of possible ingredients you can use—the size of the list merely emphasizes what a large variety of scents, textures, and effects you can incorporate into your recipes.

Ingredients

In addition to those items you may find growing in your own garden—such as many of the suggested herbs and flowers—there are many other natural ingredients that are useful for making your own beauty products. Honey, eggs, and cold-processed oils are just a few of the many you can find at any grocery store. Other, less familiar ingredients can generally be purchased at a local health food store or pharmacy.

—Flowers—

Fresh flowers are used for making your own essential oils, tinctures, and infusions. Grow your own or see your local plant nursery or florist for fresh blossoms.

Most flowers and plants are easily dried at home. Immediately after harvesting, gather the stems into small bunches and secure

Here is a listing of some of the most popular flowers used for making natural beauty products.

Flower	*Parts Used*	*Effects*	*Application*
Geranium	Flowers and leaves	Cleansing, calmative, tonic	An excellent cleanser; sometimes used in skin preparations for oily skin
Jasmine	Flowers and stems	Uplifting, balancing, acts as a tonic	Often employed in aromatherapy as an antidepressant
Lavender	Flowers and stems	Relaxing, calmative, acts as a tonic	Sometimes used in aromatherapy to help ease headaches; has some natural antiseptic properties
Lilac	Flowers	Hypnotic and relaxing	Often used as a base for creating the scent of gardenia, since the essence of gardenia cannot be captured effectively
Lily of the Valley	Flowers and stems	Hypnotic and relaxing	Antispasmodic—used in aromatherapy and some skin preparations; a favorite in lotions
Orange Blossom	Flowers	Warming and sensual; evokes a light-hearted spirit and a childlike playfulness	Aromatherapy; sometimes combined with other fragrances in body care preparations
Rose	Flowers and leaves	Relaxing and moisturizing	Commonly used in the Victorian era for toilet waters, moisturizers, and tonics; sometimes considered an aphrodisiac
Violet	Flowers	Associated with days gone by	Fragrance only; has no assigned aromatherapy values, but my testing shows it to be similar to lavender
White Ginger	Flowers and stems	Hypnotic and tropical	Used as a tropical fragrance in body care products; a favorite in Hawaii

them with small rubber bands. Don't make your bunches too large, or the lack of air flow will slow the drying time; too much moisture and insufficient ventilation will cause mold. For flowers where only the blossoms are desired, remove the leaves before bundling them. Hang the bunches upside down and allow them to remain undisturbed for several weeks in a dry, well-ventilated area. If you plan to use only the petals, spread them out on paper towels and let them dry thoroughly.

—*Herbs*—

Herbs are used in many body care products in the same way that flowers are. You can grow your own fresh herbs in the garden, in pots on your deck, or in the house. If you don't have the space or lack a green thumb, herbs are readily available from local plant nurseries, farmers' markets, and herbalist shops. Fresh herbs can be air-dried by hanging small bunches upside down in a dry, well-ventilated area. (Follow the same procedure as used for drying flowers.) Dried herbs can also be purchased at health food stores, some grocery stores, and herbalist shops.

Herbs

These herbs are commonly used in natural beauty products and are recommended for the recipes in this book.

Herbs	*Parts Used*	*Effects*	*Application*
BASIL	Stems and leaves	Antispasmodic, calming	Used in aromatherapy to help relieve headaches and stomach cramps
BORAGE	Flowers and stems	Calming, anti-spasmodic, acts as a tonic	Said to be helpful with some nervous conditions; works well for some people but can be irritating to others
CHAMOMILE	Flowers	Calming, soothing	Often included in bath preparations to soothe irritated skin
CHERVIL	Flowers and stems	Mild tonic	Used in Europe for skin problems
COMFREY	Roots	Slight antiseptic properties	Added to bath water in earlier times to promote youthful skin; currently used as an emollient for skin irritation
EUCALYPTUS	Leaves and stems	Antiseptic	Used in spas and saunas as a tonic for the lungs and to help clear breathing
GINGER	Roots	Stimulating and clarifying	Used in some hair and body products to add tone and shine
HYSSOP	Stems and leaves	Astringent and stimulant	Used as an alternative to sage in antiseptic washes
LEMONGRASS	Stems and greenery	Clarifying, acts as a tonic	Helps control oily skin, but may irritate some skin types
LEMON VERBENA	Leaves	Stimulating, acts as a tonic	Used in bath preparations for bracing and awakening

Herbs	*Parts Used*	*Effects*	*Application*
LOVAGE	Roots	Stimulating,	Used in bath preparations to soothe skin problems; cleanses and deodorizes. *(Caution: This should not be used in early stages of pregnancy — it has been used to stimulate the onset of menstruation.)*
MINT	Leaves	Clarifying, stimulating, acts as a tonic	Often used in toothpaste, mouthwashes, foot preparations. Peppermint contains menthol, which imparts a cooling sensation and acts as a natural deodorizer.
ROSEMARY	Leaves and flower tops	Stimulating, antiseptic	Common in hair preparations to help control dandruff and stimulate the scalp in hair-growth formulations
SAGE	Leaves	Astringent, clarifying, stimulating	Used as a wash and as an antiseptic on burns, bruises, and skin irritations
SAVORY	Stems and leaves	Astringent, clarifying, stimulating	Not widely used in body care products, but often considered an aphrodisiac
SOAPWORT	Roots	Mild cleanser	Used by early Native Americans as a wash (it produces a slight white foam when water is added); also used in old shampoos and body care products to help with itchy skin and skin rashes
THYME	Stems and leaves	Calming, antiseptic	Used in pure oil form (thymol) in toothpastes and mouthwashes. *(Caution: Ingesting too much of this herb can be toxic.)*

—Natural Oils—

Natural oils vary in their fragrance, thickness, and general feel. In many of the recipes, you can substitute similar oils, according to your preferences. All of these oils can be purchased at health food stores, although those that are a little less common may have to be ordered. Some, as noted below, are also available at grocery stores.

ANISE OIL: To make anise oil, seeds from the anise plant are extracted into a carrier oil. It has a distinct licorice smell and is used in products where the human smell is to be neutralized. At some health food stores, it can be found as a concentrate with the essential oils, and larger grocery stores may carry it as a carrier oil together with spices and other flavorings.

APRICOT KERNEL OIL: Taken from the large seeds (pits) of apricots, this oil is wonderful for cosmetic use. It very nearly matches the natural weight of sebum, a natural secretion of the sebaceous glands. Your skin will drink up this oil, which is very nourishing to the skin.

BORAGE SEED OIL: Unlike many oils, which are expressed from seeds or nuts, this is made by extracting the essence of the plant into a carrier oil. Because of its calming properties, small amounts of this oil are generally used as a tonic. It has a soothing, anti-inflammatory action. Borage seed oil is not easy to find, but most health food stores can order it for you. It is available as a concentrate/essential oil and as a carrier oil.

CALENDULA OIL: Like borage seed oil, this is not an expressed oil, but one that is created from the plant and concentrated in a carrier oil. It has calming and soothing properties and is often used in baby formulas. It can be found in concentrated form and as a carrier oil.

CANOLA OIL: This oil comes from the Canadian plant *brassica*

napus, or *brassica campestris*, which is a cousin of the rapeseed plant and has been specifically altered to produce low levels of erucic acid. It is a light, edible oil.

CITRONELLA OIL: Made from the leaves, stems, and flowering tops of a tropical Eurasian plant, this has long been praised for its ability to repel insects. Frequently used in candles for the garden, it can also be used in skin preparations for warding off insects. (Pennyroyal has a similar effect, but it can be toxic.) Look for the concentrate in the essential oil section at your local health food store. It is also available as a carrier oil.

COCONUT OIL: When you see this natural oil on a cool day, you may be surprised to find that it is white and solid. It liquefies easily with warmth and will not solidify again at room temperature if it is mixed with other oils. Hawaiians commonly use this oil to moisturize their skin and add shine to their hair. It's used in formulations for treating skin that has been exposed to the sun and is a popular ingredient in massage oils.

EVENING PRIMROSE OIL: Too expensive to use in large quantities, this oil is extracted in the same manner used to make borage seed oil. It is used as a soft astringent and to relieve skin irritations. Recently it has been recognized as an aid to soothe some premenstrual symptoms. It is available as a concentrate/essential oil and as a carrier oil.

GRAPESEED OIL: A light oil, without too much odor, this has a fair shelf life. Some people prefer to use this in place of sweet almond oil.

HAZELNUT OIL: Although it has a relatively brief shelf life, this oil is a nice emollient that is suitable for the skin or hair. It is expressed from the nut.

Grapeseed

Sweet Almond

Safflower

Apricot Kernal

JOJOBA OIL (GEL): Juice from the jojoba plant is usually available in some type of carrier oil or as a gel. Jojoba has a waxlike consistency that is ideal for adding body to a mixture of oils, and it is very effective when used alone. Often used as an emulsifier in skin care preparations, it nourishes the skin. Jojoba has an extended shelf life. It can sometimes be found in drug stores.

MACADAMIA NUT OIL: Taken from the nut, this oil is a rich emollient that works well in skin and hair preparations.

OLIVE OIL: Made from pressed olives, this oil is often used as a base for suspending other ingredients. It is used on hair and skin as an emollient, and when combined with vitamin E, it is said to be beneficial in healing skin burns. Cold-pressed, extra-virgin oil is recommended; it can be found in most grocery stores.

SAFFLOWER OIL: Used frequently in massage products, this oil has a nice weight and an acceptable shelf life. It is readily available in grocery stores.

SUNFLOWER OIL: A light, pleasant oil, this has some natural odor. It is better suited for the bath than for massage and is similar in usage to canola oil. It can often be found in grocery stores.

SWEET ALMOND OIL: Expressed from the nut, this is a multipurpose oil that is great for cosmetic use. It's used in many bath and skin care formulas because it is a good emollient. It has a moderate shelf life and has very little natural smell.

WHEAT GERM OIL: High in vitamins A, D, and E, this oil also contains lecithin. It has a longer shelf life than some oils, but it sometimes has a strong odor.

—Other Ingredients—

These ingredients can all be found in your local health food store. As noted below, some of the more common ones are also available in drug or grocery stores.

ALOE VERA GEL: The clear gelatinous liquid taken from the aloe plant has been used since antiquity for its healing effects on burns, sores, and skin irritations. It soothes and softens the skin and helps burned tissue regenerate without scarring. When applied by itself, aloe slightly tightens the skin and is somewhat drying. Look for pure aloe vera gel that has no additives.

ARROWROOT POWDER: This is a starch from the root of the arrowroot plant (so called because it was once used by Native Americans to heal wounds from poison arrows). It is similar in its properties and uses to cornstarch but is somewhat finer—its consistency is between that of cornstarch and rice flour. It has wonderful drying abilities.

BAKING SODA: Also called bicarbonate of soda, this is a natural odor-eater that also has skin-soothing properties and a mild bleaching effect. Its light abrasiveness and bleaching action make it an effective ingredient in some toothpastes. It is also used in some bath preparations for soothing irritated skin. My mother used to make a paste with water and baking soda to put on our bee stings, and it helped soothe the inflammation. Baking soda is sold in any grocery store.

BEESWAX: In beauty products, beeswax is used as a waxy emollient and natural emulsifier. It also forms the base for some cosmetics. Local bee keepers are potential sources of beeswax.

BENZOIN: This whitish resin, taken from any of several varieties of the styrax tree, was used in Victorian formulas to help *fix* or keep a fragrance longer. It is still used as a fixative in some skin care products, but benzoin by itself can be irritating.

BORAX: This mineral is used as a stabilizer and drying agent in some cosmetics. It may be a bit of a challenge to find it, but your local health food store can help you.

CAMPHOR: This volatile compound is extracted from the wood and bark of camphor trees or produced synthetically. It was used in early medicines for lung and breathing problems, and it stimulates the clearing of nasal passages when a cold is present. Camphor is a tonic and is also used in sports rubs.

CASTILE SOAP: Named for the region in Spain where it originated, this is a very mild soap made with olive oil. Liquid castile soap is used as a foaming base for some products. It can be found in some drug stores.

CITRUS PEELS: The rind from oranges, lemons, grapefruits, limes, tangerines, and other citrus fruits contains significant amounts of vitamin C, which is noted for its anti-aging properties in skin care products. Separate the peels and chop them with a knife or use a food processor to cut them into small pieces (the finer the cut, the better). Use the fresh pieces right away if you're making essential oils; otherwise, let them dry and store them in a sealed container until you're ready to use them.

COCOA BUTTER: Available in bar form, this slightly yellowish fat is made from cacao beans, which accounts for its chocolatelike scent. It melts at a very low temperature and generally becomes liquid when it comes in contact with the skin. It is used as a soothing, moisturizing ingredient in a variety of skin care products.

CORNSTARCH: A natural powder that is used as a thickening agent in cooking, cornstarch is included in body care products to improve absorption, to help soften the effects of acidic ingredients, and to increase viscosity. It is noted for its cooling, drying, and smoothing properties and is frequently used as a base for body powder in natural formulas. Cornstarch is available in the baking section of grocery stores.

DEAD SEA SALTS: These salts originate from the area around the Dead Sea. Although they produce the same effect as standard sea salts, Dead Sea salts have a higher mineral content and are thought to be more beneficial. Their mineral content can vary, however, according to where the salts were harvested. These salts can be more difficult to obtain but are available at some health food stores.

DISTILLED WATER: Unlike tap water, distilled water contains no bacteria, minerals, or chlorine. These unwanted additives can affect your formulas in unpredictable ways. To be on the safe side, always use distilled water in the recipes that call for water. It's available in any grocery store.

EPSOM SALTS: Originally obtained from the mineral springs in Epsom, England, these clear or white crystals have long been recognized as an effective anti-inflammatory. Soaking in epsom salts helps reduce swelling and inflammation, and it softens the skin. Epsom salts are available in drug and grocery stores.

ETHYL ALCOHOL: Derived from grain, this type of alcohol is the intoxicating ingredient in liquor, and in its pure form, it has very little odor. Grain alcohol can be purchased under a variety of trade names at liquor stores. Ethyl alcohol is not to be confused with *methyl* alcohol, which is made from wood and should *not* be used in your natural beauty products. It can be fatal if taken internally, and repeated inhalation of the fumes from methyl alcohol can cause injury to the nervous system. Vodka, an alcoholic liquor made from potatoes, rye, or corn, can be substituted for ethyl alcohol. To get an adequate alcohol content, purchase 90-proof vodka. *Isopropyl* alcohol (also known as rubbing alcohol) can also be substituted for ethyl alcohol, but it has a significant odor. It can be purchased at pharmacies and grocery stores.

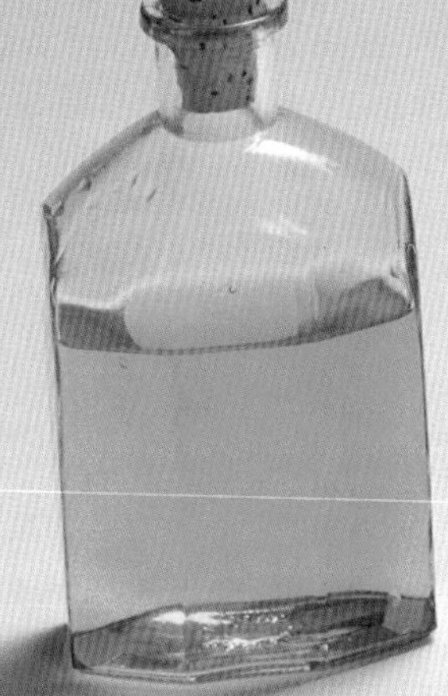

Aloe Vera Gel

Glycerine

Vinegar

Epsom Salts

Citrus Peels

Menthol

Beeswax

FOOD COLORS: The U.S. Federal Drug Administration has cleared certain colorants for use internally. You can buy these at the grocery store and use them to color any of your bath and beauty products that are water based. They are not suitable for coloring oil-based products, however. Food colors can be blended to make many shades, but all should be used in small amounts. If used in too great a concentration, they can stain the skin temporarily. Note also that some people are sensitive to colorants. Food colors are available at your local grocery store.

FRENCH CLAY: This fine-quality natural clay is primarily used in facial masks for tightening pores and absorbing oil from the skin. If you can't find French clay, a suitable alternative is kaolin, a fine, white clay (also called China clay) that has an astringent, cleansing action.

FULLER'S EARTH: White to brown in color, this naturally occurring claylike substance has a stimulating and cleansing action when used in facial masks. In other preparations, it can be used as a binding agent. Most health food stores can get it for you if they don't stock it.

GLYCERINE: Rendered either from animal fats or vegetable oils, glycerine is a "super-emollient" that separates itself as cream does from milk. It is skimmed off and saved to use as a moisturizing ingredient. Glycerine is a humectant—a substance that promotes moisture retention—and soap made with glycerine sometimes develops water droplets on its surface from the moisture it has drawn from the air. Glycerine is sold in different grades, depending on its quality. It can be found at some drug stores.

HONEY: This natural product of bees is most commonly used as an emollient or moisturizer in beauty preparations. In small amounts, it is soluble in water or oil, although it must be heated and stirred into oil for it to emulsify. It is easily found in grocery and health food stores.

LECITHIN: Yellow-brown in color, this waxy emulsifier has a consistency similar to thick honey. It can be obtained from egg yolks and is a natural by-product of the manufacture of soybean oil.

LOOFA: Creamy off-white in color and cylindrical in shape, a loofa is the dried, fibrous portion of the luffa fruit, a tropical vine. Also called the "dishcloth gourd," it is primarily used to exfoliate the skin. Loofas in many sizes and shapes can be found in the bath section of most department and discount stores.

LYE: This is a strong alkaline solution that was originally made by leaching wood ashes with water. It is a highly caustic substance—it's the main ingredient in drain cleaner—and must be handled with care. It is used in cosmetics only for soap making. When the oils and lye are mixed together at the proper temperature, the chemical reaction produces soap. Lye is available at hardware stores.

MENTHOL: A crystallized concentrate of mint, menthol has traditionally been used as a therapeutic ingredient in sports rubs and foot care items. When it comes in contact with the skin, it causes a pleasant tingling and warming/cooling sensation. It can be hazardous if used in too high a dosage.

PARAFFIN: This white or translucent, odorless wax is obtained from petroleum distillates. It's mainly used to seal the corks on your bottles of massage oils and other liquid products. You can find paraffin at your local craft store in the candle department or at grocery stores that sell canning supplies.

PETROLEUM JELLY: This is a by-product of petroleum distillation and isn't "natural" by my strict definition. Some health food stores are now featuring a nonpetroleum jelly product that makes an effective substitute; beeswax is another alternative. Petroleum jelly provides a thick but spreadable consistency to skin care products and is commonly used commercially. It is available at drug and grocery stores.

POWDERED MILK: Milk is soothing and softening to the skin, and it can be used for skin bleaching. Powdered milk provides all the same benefits and is used when a liquid ingredient is not desirable. It is readily available at grocery stores.

PUMICE: White, gray, or brownish-black in color, pumice is a very lightweight volcanic glass. Its characteristic rough texture is produced by the presence of water vapor in the molten lava. Pumice is used for buffing rough or calloused skin. Ground pumice is not easy to find but can be ordered from some pharmacies.

RICE FLOUR: Similar to cornstarch in its characteristics and uses, rice flour is a much finer powder. (It is also finer than arrowroot powder.) The geishas of Japan used this fine white flour on their faces, and you will find that it makes a lovely cosmetic powder.

ROSEWATER: True rosewater is the fragrant by-product of distilling rose essential oil. You can approximate it by making an infusion of rose petals (see page 36) or by adding pure rose essence to distilled water. Rosewater was used in Victorian days as a beauty treatment to soothe the skin. It is still used today in skin formulas where a rose scent is desired.

SALICYLIC ACID: This white crystalline compound is more commonly known as aspirin. The tablets can be crushed, and when diluted with water, this acid works very well as an antiseptic and helps control dandruff. Be sure to buy pure aspirin, not one of the analgesic blends. It's available at drug and grocery stores.

SAND: Very fine sand is used for exfoliation—the removal of dead skin cells—especially on the feet. If you can't find it anyplace else, try an aquarium shop or building-supply store (look for play sand, which is finer than builder's sand).

SEA SALT: Dried salt from the sea has been used as a toner and makes a wonderful scrub and tonic in your bath water. It also has water-softening abilities. Salt is a bit dehydrating, though, so make sure you follow its use with a glass of water and your favorite moisturizer. If you can't find it at your local health food store, inquire at stores carrying aquarium supplies.

SOAP FLAKES: These are powdered bits of any simple lard-and-lye variety of soap. They can be difficult to find, but they are carried by some old-fashioned drug stores and some specialty stores, or you can order them from a pharmacy or health food store. Otherwise, you can make your own by grating a bar of your favorite soap.

TEA TREE EXTRACT AND OIL: Extracted from the melaleuca, a tree native to Australia, this substance has been found to have amazing antiseptic properties. Its antibacterial and antifungal properties make it a valuable addition to many preparations on the market. Two of the most popular are foot formulas and antiseptic creams. Tea tree has a strong odor but a good shelf life. The extract is less commonly available than the oil but is stronger in concentration; be sure to note which is specified in the recipe you're using because the required amounts will differ if you substitute one for the other.

VITAMIN E OIL: This is vitamin E suspended in a carrier oil. Read the label carefully to determine its concentration, and be sure to choose the best quality. Vitamin E has been shown to have great moisturizing and nourishing effects on the skin. Because of its superior healing qualities, medical centers use it on the skin of burn victims to promote healing. Vitamin E is a natural preservative and will help sustain the shelf life of your formulas.

WHITE VINEGAR: Vinegar is made by fermenting fruit or wine and comes in a number of interesting varieties. Some have considerably less odor than others, and some are flavored (and scented) by herbs held in suspension. Vinegar is a superb toner because of its acid content, and it is not as harsh as alcohol. It's frequently used in tonic baths and for bracing the skin. Most grocery stores carry various types of white vinegar.

WINTERGREEN: Taken from the wintergreen plant, this is almost always presented in an oil. In the old days it was added to disagreeable medicines to disguise the taste. It is still used in many forms and is most often combined with menthol or peppermint in foot care products and sports rubs.

WITCH HAZEL: Liquid derived from the leaves and bark of this winter-blossoming shrub is often used in skin and hair preparations. It can be applied directly on skin irritations, such as insect bites and scrapes. Known for its cleansing properties, witch hazel is a good astringent and is recommended for use on oily skin or acne. Drug stores often carry it.

Supplies & Equipment

In addition to your materials, you'll need a few simple tools and other supplies:

- A reliable kitchen scale
- Candy thermometers (two are needed to make your own soap)
- Eye dropper
- Measuring spoons
- Glass measuring cups
- Mortar and pestle
- Clean mixing utensils made of stainless steel or other metal, not plastic or wood
- Glass, ceramic, or stainless steel mixing bowls
- Large plastic tubs with lids for storing dried herbs or flowers
- Stainless steel or glass double boiler for stove top
- Stainless steel, glass, or enamel pots for stove top
- Plastic candy or cookie molds
- Glass jars with lids, such as canning or mayonnaise-type jars
- Decorative or unusual jars, tins, and bottles

Keep in mind that you also may want to accumulate fabric, lace, ribbons, and dried and silk flowers for decorating your containers. You may want to add small labels too.

1 QUART
32 OZ
4 CUPS
3 CUPS
1 LITRE
¾ LITRE
½ LITRE
¼ LITRE
PYREX
1 CUP
8oz
225
175
125
75
METRIC

Chapter Three

Basic Preparations

MAKING YOUR OWN NATURAL beauty products is surprisingly easy, and most of the ingredients you'll use are readily available in your local health food store, pharmacy, or even the grocery store. A few of the substances you'll need, such as essential oils, tinctures, and soap flakes, are a little more complicated and can be purchased as you need them. If you like making all your preparations from scratch, this chapter shows you how.

Techniques for Gathering Fragrances

Originally all fragrance compounds were made from organic materials, such as flowers, spices, and resins. Through the years as the technology developed, it became evident that major fragrance manufacturers were totally dependent on the success or failure of certain crops grown for perfumery. Supply and demand pressures forced the development of synthetic essences, which are now widely available and less expensive than their natural counterparts.

Synthetic compounds include manufactured scents and reproductions of natural scents, and they're produced by mixing the appropriate chemical constituents, such as polymers, terpenes, aldehydes, ketones, and alcohols.

Natural fragrances are derived from living tissues, mainly plants. Many aromatic plants have tiny sacs that store a fragrant substance. This substance is an oil, which perfumers call an *essence*. Perfumers extract the essential oil from the flower blossoms, leaves, stems, or roots of plant materials usually by one of the three following methods: distillation, expression, and enfleurage.

DISTILLATION is a method in which the plant material is boiled in water or subjected to a concentration of steam, which releases the essential oil into the vapor. The aromatic gas is then passed through glass tubing, where it is cooled and condensed into a liquid.

EXPRESSION is the process of pressing the oil out of the plant tissue. Sometimes the essential oil is found in the rinds of fruit, such as the lemon or orange, or in pulpy plant material. If this is the case, then expression is generally the easiest method for obtaining it.

ENFLEURAGE AND EXTRACTION is an age-old way of releasing the essential oils from flowers, grasses, or plant matter. Freshly picked flower petals are spread over glass plates that have been covered with fat. The essential oil from the petals is absorbed into the fat, where it is allowed to mature. Then the fragrance is extracted by placing the fragrant fat into a closed container with alcohol and applying heat. The alcohol dissolves the essential oil and rises with it to the top of the liquid fat, where it is then skimmed off. Using enfleurage and extraction, one ton (907 kg) of flower petals produces only about 10 to 16 ounces (296 to 473 ml) of essential oil.

MACERATION is very similar to enfleurage and extraction. The only difference is that the container of oil or fat is placed in hot water for a few hours every day to speed the absorption process.

Animal Fragrances

Scents that are animal in origin are used in some perfumes. Among those that can be used, these are the most common:

CASTOR: a creamy, orange-brown substance with a strong, penetrating odor, obtained from the perineal glands of a beaver

CIVET: a fatty substance with a strong, musky odor that is found in a pouch near the sexual organs of the civet cat, a small African mammal

MUSK: an oily liquid formed in a sac under the skin in the abdominal area of the male musk deer

Animal fragrances are considered valuable because they are strong, penetrating, and long lasting. In large quantities animal-derived essences are sickening, but small amounts added as tinctures to the scent base provide a musky or muted animal undertone that some find desirable. There are only two methods for achieving these scents: synthetic reproduction and deriving the essence from the animal—let your conscience be your guide.

Essential Oils

Essential oils are used for aromatherapy purposes, scenting bath and body preparations, refreshing or making potpourri, and blending into perfumes or colognes. Many of the recipes in this book include essential oils, and it's fun to try making your own at home. There is just one note of caution: Although you will be making your essential oils from natural ingredients, please do not be tempted to take them internally. Essential oils are very strong and often are not safe to consume. Additionally, your plants may not have been prepared in sterile conditions.

True essential oils are extracted from only one scent, but you can create your own mixtures of fragrances by extracting them from several flowers and herbs together. Such a blend may smell wonderful or not good at all—you may want to save this alternative until you're a little more experienced with what goes together well. Listed below are some flowers, herbs, and spices that are suitable for making your own essential oils.

FLOWERS

Carnation	Lilac
Geranium	Lily of the Valley
Hawthorn	Orange Blossom
Heliotrope	Rose
Honeysuckle	Rose Geranium
Hyacinth	Sweet Pea
Jasmine	Violet
Lavender	White Ginger

CITRUS AND SPICES

Cinnamon bark	Lime peel
Cloves	Orange peel
Grapefruit peel	Tangerine peel
Lemon peel	Vanilla beans

HERBS

Basil	Lemongrass
Borage	Lovage
Chamomile	Mint
Chervil	Rosemary
Comfrey	Sage
Ginger	Savory
Hyssop	Thyme
Lemon Balm	Verbena

The easiest way to make essential oils at home is to use a simple variation of the enfleurage and maceration techniques. Fresh herbs and flowers make the strongest essential oils; however, if you do not grow your own or have access to fresh plants, you can use dried products instead.

—*Floral Essential Oils*—

To make an essential oil from flowers, you will need ½ cup (118 ml) of cold-pressed sweet almond oil and 2 to 3 cups (473 to 710 ml), packed, of freshly cut, full-bloom flowers. Try to cut the flowers in the morning, when they are freshest, and be sure they have not been treated with pesticides.

Put the oil in a glass canning jar or other wide-mouth glass bottle, and use a mortar and pestle to bruise the flowers slightly. Then add the flowers to the oil. Cap the jar and shake it thoroughly to bruise the flowers further. Leave the jar in a moderately warm place for a week and shake it every 24 hours. If the ambient temperature is below 68 to 75°F (20 to 24°C), then place the jar in a warm-water bath once every 24 hours.

Put the jar in a dark place and let it age for another one to two weeks so that the oil can fully absorb the essence of the flowers. Then bring it out and shake it again. Strain the oil through a piece of fine-mesh cotton gauze, gathering the flowers into the gauze as if it were a pouch. Squeeze or mash the gauze pouch to get as much of the floral essence into the oil as possible. Discard the flower pulp and gauze. Transfer your oil to a dark-colored glass bottle and keep it airtight with a lid or stopper, storing it between uses in a cool, dark place. The shelf life is 12 to 18 months if stored away from sunlight and kept at temperatures lower than 75°F (24°C).

—Herbal Essential Oils—

To make an herbal essential oil, you will need ½ cup (118 ml) of cold-pressed sweet almond oil and 1 cup (237 ml) of fresh herbs. As before, make sure the herbs are well packed into your measuring container. Then follow the same procedure as for flowers to make the essential oil.

—Spices & Citrus—

An essential oil made with citrus requires ½ cup (118 ml) of cold-pressed sweet almond oil and 1 cup (237 ml) of citrus peel that has been chopped into small pieces. When using spices, you will need ½ cup (118 ml) of sweet almond oil and ½ cup (118 ml) of dried spice pieces, not powder. Follow the same procedure used for floral essential oils.

Tinctures

Unlike essential oils, tinctures rely upon alcohol to coax the essence from a flower or herb. Fresh plant material is bruised—lightly crushed with a mortar and pestle—and placed into the alcohol to steep. After a period of time, the essence is released into the liquid. (A related compound is an extract, which also uses alcohol but is made with *dried* plant materials.) Through the ages, herbal tinctures have been commonly used for their healing properties. Tinctures also make effective ingredients in some creams and ointments, and they can be added to the bath. Any of the flowers, herbs, or citrus and spices listed on page 32 are suitable for making tinctures.

The customary method for making a tincture is to use 1 part of herb or plant material and 5 parts of alcohol. You should not try to concentrate the solution by increasing the amount of flowers or herbs or by using less alcohol because that increases the risk of bacteria growth.

Use either ethyl alcohol (pure grain alcohol) or 90-proof vodka for making tinctures. When purchasing a vodka, choose one with the least amount of odor so that it won't unduly influence the scent of your tinctures.

For a typical tincture, you will need 1 cup (237 ml) of plant matter, tightly packed to give a full measure, and 5 cups of alcohol. Using a mortar and pestle, lightly bruise the chosen plants and place them into a glass jar. Add the alcohol, seal the jar, and place it in a dark, warm location. Shake the jar twice a day for a total of 14 days. Using a piece of fine-mesh cotton gauze, strain the liquid into a clean jar, firmly squeezing the cloth to get the full concentration. Store the tincture in a sealed, dark-colored glass bottle in a cool, dark location. Its shelf life is 18 months to two years.

Although you're using natural ingredients—and the spirit content is inviting—don't be tempted to ingest your tinctures. These substances are very concentrated and should be treated with respect. Use them as ingredients in recipes for products that will be applied to the skin or added to the bath.

Herbal & Floral Infusions

Like making tea, infusions are prepared by pouring hot water over specific flowers or herbs and allowing them to steep. The essence of the plant is infused into the water and makes a lovely, though short-lived, preparation to add to your natural beauty products. Infusions make wonderful bath "teas" to use within a few days. You can extend the shelf life by nearly a month if you add a small amount of alcohol to your infusion after it has cooled to room temperature.

Infusions can be made from a single ingredient, or you can combine a few plants for a fragrant mixture. Any of the herbs and flowers listed in the section on essential oils will work for infusions. Spices don't generally infuse well, however; they often cloud the liquid. (If you don't mind the cloudy appearance, a spice infusion can be made by adding 1 percent alcohol to the mixture.)

Two of my favorite combination infusions are chamomile/lavender and geranium/rose. For the first you will need ½ cup (118 ml) each of fresh chamomile and fresh lavender flowers; for the other, use ½ cup (118 ml) of fresh geranium flowers and ½ cup (118 ml) of fresh rose petals.

Put the fresh flowers together in a glass bowl and pour 2 cups (473 ml) of boiling water over them. Cover the bowl and let the flowers steep for 15 to 20 minutes. Using a piece of finely woven cotton gauze or cheesecloth, strain the liquid and transfer it into decorative bottles. Remember to use any infusions within one week. You can stretch this to almost four weeks if you add 1 tablespoon (15 ml) of alcohol to the cooled infusion and shake it well.

Rosewater is one of the most popular infusions, and it is used in several of the recipes in this book. To make your own, use a cupful of freshly picked rose petals and follow the instructions above.

Decoctions

Decoctions are similar to infusions, but the heat is applied more directly and intensely. To make a decoction, the plant matter is placed in a pan of water, covered, brought to a boil, and simmered for a period of time until the essence of the plant enters the liquid. This process is especially effective for nuts or woody plants, which do not release their essences as readily as soft leaves and petals.

For a typical decoction, place ½ cup (118 ml) of crushed nuts and 2½ cups (591 ml) of water in a small saucepan. Cover and simmer it for 15 to 20 minutes. Using a slotted spoon or potato masher, mash the nuts and simmer for another 5 minutes. Then strain the liquid into a decorative bottle and discard the pulp.

Handmade Soap

Making your own soap is much more complicated than creating essential oils or brewing herbal infusions, and there are many fine books on the market covering the traditional methods. However, a few of the recipes in this book call for soap flakes, and it's enjoyable to be able to make all your beauty preparations from scratch if you so desire. Here are two less traditional soap formulas that do not contain animal by-products or alcohol.

—Natural "Veggie" Soap—

This soap is naturally light ocher-amber in color and is an updated variation of an old Italian soap formula. It is very mild and not as drying as regular lard-and-lye formulas. It's perfect as an uncolored and unscented bar, but you can also add color and fragrance, if you wish. This formula will not be as translucent as many of the commercial glycerine soaps, some of which use alcohol to obtain their clarity. You will need the following ingredients:

6 cups (1.4 l) olive oil
2¼ cups (532 ml) coconut oil
3½ cups (828 ml) solid vegetable shortening
1½ cups (354 ml) lye
4¼ cups (1 l) cold water

In addition, you will need two candy thermometers—one for the oil and shortening combination and another for the lye and water.

Slowly add the lye 1 tablespoon (15 ml) at a time to the cold water while stirring continuously. *(Caution: Lye is very caustic and should not come in contact with your skin. This is not a procedure for children!)* It's best to wear rubber gloves and use a mask or avert your face while stirring the lye solution to avoid inhaling any fumes. The chemical reaction of the lye and water produces considerable heat. After you have stirred thoroughly, allow the mixture to sit until the temperature drops to between 95 and 98°F (35 and 37°C).

As the lye mixture cools, place the olive oil, coconut oil, and vegetable shortening in a pot, preferably stainless steel, and heat it to 125 to 130°F (52 to 54°C). Remove the pan from the heat, place the thermometer in the liquid, and let the mixture cool, watching the temperature.

The temperature of both solutions must be either 96 or 97°F (35.5 or 36°C) at the same time in order for you to make soap. (This is the real trick of soap making—and the hardest part.) When this happens, stir the oil mixture for about 30 seconds; then add the lye mixture slowly but evenly in a steady stream. Stir continuously while adding the lye solution and keep stirring until you think your arm will fall off! Alternate stirring hands if necessary, but do not stop. When the consistency is like a very thick gravy or sauce, the soap is ready for you to add a few drops of food color and ¼ teaspoon (1.25 ml) of essential oil, if desired. Continue stirring when adding color or scent; then pour the liquid soap into one or more molds.

The simplest type of mold is a wax-coated cardboard milk container. You can cut it to the desired length, and when the soap sets up, you can tear away the sides to release the soap.

To make bars of soap, pour the solution into a rectangular flat pan large enough to make the soap ½ to 2 inches (1.3 to 5 cm) thick. When the soap has had 24 hours or more to set up, use a hot knife to cut it into squares and remove it from the pan. You may need to buff the edges with a soft cloth.

My favorite technique is to use a plastic pan that is thick enough to hold the heat without melting, yet flexible enough to be bent. With this type of pan, you can unmold the whole block of soap at once by twisting and pressing the pan. Then put on heavy gloves or cooking mitts, heat a strand of wire about 12 inches (30.5 cm) longer than the length of your soap block, and use the hot wire to cut your soap into bars. Very little buffing will be needed.

To make decorative soaps, you can use plastic molds—candy molds are best—or you can make soap balls. During the set-up process, when the soap is pretty thick but not yet hard, use an ice cream scoop to dig out round balls that can be smoothed into shape with your hands. Set them on waxed paper to harden.

After molding or shaping your soap, let it sit for two to three days in a dry place to cure. Once it has cured, this vegetable soap does better if wrapped. This type of soap is so full of moisture that it tends to develop water beads if left in the open air too long without use.

—Basic Glycerine Soap—

The warm, honeylike color and translucent appearance of glycerine soap make it especially appealing. If you prefer a low-sudsing soap, omit the optional commercial shampoo from the formula. It's included only to provide more bubbles, not to add cleansing action. These are the ingredients you'll need:

4 heaping tablespoons (about 60 ml) lye
1 cup (237 ml) cold soft water
⅔ cup (158 ml) glycerine, preferably vegetable glycerine, but animal-derived glycerine will also work
3 tablespoons plus 2 teaspoons (54 ml) coconut oil
2 cups (473 ml) lukewarm water
2 tablespoons (30 ml) commercial shampoo (optional)

While gently stirring, slowly add the lye 1 tablespoon (15 ml) at a time to the cold water. As noted in the instructions for the vegetable soap, handle the lye with care. Wear rubber gloves and a mask and stir slowly to prevent any spattering. After you have stirred thoroughly, allow the mixture to rest until the temperature drops to about 90 to 95°F (32 to 35°C).

Meanwhile, heat the glycerine and coconut oil in a pan until the mixture reaches 150°F (65°C). When both mixtures are at the correct temperature, slowly pour the glycerine and oil mixture into the dissolved lye. Stir constantly until the combination has thickened into a honeylike consistency. This will take about half an hour, depending on the outside temperature and other factors. Then pour the liquid soap into molds and allow it to set up as described for the vegetable soap. After it has set up and been cut into bars, let the soap cure for one to three days.

—*Adding Variations*—

Try adding small amounts of the following ingredients to your soap batches to create a variety of types of soap.

- Almond meal, oatmeal, or cornmeal to make a scrub soap
- 2 tablespoons (30 ml) of rosewater and 6 drops of rose essential oil for a fragrant rosewater soap
- Herbs, such as mint, camphor, eucalyptus, or rosemary
- Cedar or balsam to make a soap that especially appeals to men
- Anise oil for a fisherman's soap
- 2 tablespoons (30 ml) of powdered milk and 1 teaspoon (5 ml) of honey for a milk and honey soap
- Dried fruits or vegetables, such as strawberry or cucumber
- Citronella and thyme to make a soap that works as an insect repellent—a perfect gardener's soap
- Spices, such as cinnamon and cloves, make wonderful aromatic soap. These can be added as essential oils or as small chopped pieces.

Many of my students have been tempted to add fresh fruit for fragrance, but this doesn't work very well due to spoilage. Add only dried fruits, and if you do, try a small batch first. This will allow you to see if you like the look and effect of the dried fruit and make sure that it sets up and cuts well. Check your test batch after two to four weeks; if it hasn't developed an off color or unpleasant smell by then, it will probably be fine and have an acceptable shelf life.

—*Troubleshooting*—

- If your soap doesn't set up and separates, it probably wasn't hot enough or both solutions were not the correct temperature when you added them together.

- If your soap looks uneven or lumpy after it has set up, there are several possible causes: the amounts of your ingredients were not correct, the soap cooled too quickly, you added the lye solution too slowly, or possibly your stirring was too slow or too fast. Soap making is temperamental, but once you get your technique established, it is very rewarding. Sometimes it's perfect the first time; sometimes you need to do a few batches before you can develop your style.

A quick and easy alternative to making soap from scratch is to save soap scraps and melt them all together. By adding your favorite essential oils, dried herbs, or other ingredients, you can still create custom soap. (Be sure to add no more than 1 percent of herbs or other ingredients by volume.) You can also purchase commercial soap bases, which can be melted, scented, and molded into the desired shapes.

To Color or Not to Color...

Most of the beauty products that you'll make from the recipes in this book can be colored if you wish. Basic food colors that are available in your grocery store will work for all water-based formulas, such as lotion, cream, soap, after-shave, and shampoo. Not recommended for coloring are any oil-based products, such as essential oil, perfume, cologne, massage oil, and bath or body oil, which require special commercial oil-based colorants not available to consumers. Candle colorants, paint pigments, and dyes are not always safe for use on skin, so resist any temptation to use them.

Do not attempt to color dusting powders with food colors; these will clump unattractively, and you will not be successful. Facial treatments, such as masks and eye creams are not appropriate for coloring because of the sensitivity of the skin in these areas. Products intended for use on a baby's skin also should not be colored because some babies are sensitive to coloring agents.

Add just a few drops of color at a time until the desired shade is achieved. A little bit of color goes a long way, and too much may leave a residue. It's best to keep your products pastel and never add more than 1 to 3 percent of the total volume of the product. Be wary of coloring your products very pastel pink or purple, as they can fade and alter color easily.

Most food colors are available only in basic hues, such as red, blue, green, and yellow, but these can be mixed to obtain numerous other shades. Here are some sample blends.

- Teal or aqua: Add a small amount of blue to green
- Pink: Use a small amount of red only
- Coral: Add a small amount of yellow to pink
- Orange: Combine equal amounts of yellow and red
- Burgundy: Add a small amount of red to purple. (This shade is more light-stable than true purple.)
- Purple: Combine equal amounts of red and blue

Chapter Four

Aromatherapy

FRAGRANCES DERIVED FROM RESINS and plant materials have been used throughout the ages for healing the body and soothing the mind. Before more sophisticated methods were developed, early peoples burned branches that had been dipped in tinctures or plant extracts and inhaled the smoke to purify the body and protect it from disease. By trial and error, herbalists throughout the world and in various cultures developed their knowledge of the healing properties and the effects on mood of various herbs and flowers.

Aromatherapy Techniques

In recent years, there has been a strong resurgence in interest in natural healing practices, and the stress of today's fast pace has led many to seek gentle, natural methods for relaxation and mood enhancement. Aromatherapy offers benefits in both areas. Today's practitioners use three basic methods for administering aromatherapy: inhalation, topical application, and ingestion.

INHALATION: With this method, the plant extract or essence is released into the air and inhaled. It is generally done with the assistance of gentle heat, often provided by a small candle or light bulb. This technique is most commonly applied to evoke specific moods or feelings and with memory association. It is also used by some professional aromatherapy practitioners to promote healing.

TOPICAL APPLICATION: Contemporary aromatherapists apply essential oils to the skin for healing purposes. Because essential oils are readily absorbed into the skin and actually enter into the bloodstream, their effects are not necessarily confined to the area where they were applied.

INGESTION: Placement of small amounts of essential oil under the tongue is done solely for medicinal purposes.

This book addresses only the inhalation method and its use for evoking particular moods, feelings, and memories. Healing with aromatherapy is best left to professionals who have extensive experience in this field. Plant extracts and essences are the sources of many potent medications, and despite the fact that they are natural substances, essential oils can bring more harm than benefit if applied incorrectly or in the wrong amounts. If you are ill, consult with your doctor first.

As we explore some of the more common applications for organically derived essential oils used in the inhalation method, keep in mind that the responses noted are taken from numerous control groups that have been tested over the years by researchers. Although these are the most prevalent responses, that doesn't mean there aren't exceptions or that you will feel the same effect.

The one factor that can't be controlled from one person to the next is memory. Here is a radical example: Suppose a person had been brutally robbed in an area where jasmine was blooming in abundance. Although that person may not consciously recall the presence of the flowers during the assault, he or she would most likely have a starkly

unpleasant response to the scent of jasmine, which—according to research—usually evokes calm and alleviates depression.

Aromatherapy is a delightful science but not an easy one. In addition to observing time-honored aromatherapy responses, sample a few of your own memory associations. Close your eyes and smell ground cinnamon. What does it bring to mind? The most common responses are: 1) baking and 2) Christmas. Now try this with a strong gardenia perfume or cologne. With this scent the most frequent associations are: 1) high school prom; 2) first date; and 3) spring/summer in the South.

Creating Aromatherapy Oils

Essential oils are commonly used in aromatherapy today, and for inhalation purposes, an essential oil efficiently captures the aromatic essence of the plant, herb, or spice desired. You can create essential oils from the appropriate plants or flowers by following the procedures described on pages 31 to 33, or you can purchase some of these essential oils at your local health food store. (Just remember that you are shopping for an organically derived, potent substance, not a perfume. Don't buy scents that are produced artificially.)

Historically, aromatherapy employed just one essence to achieve the desired effect. Today aromatherapists frequently combine essences with "like values" to create synergy, thus amplifying the effects.

A few of my favorite blends are given here for you to try. Unlike the other recipes in this book, the amount required for each ingredient is described as a percentage, not an exact amount. Depending upon how much oil you wish to make, you can adjust the unit of measurement accordingly—just make sure to keep the same percentages.

—Relaxing Blends—

80% lavender, 15% rose geranium (or other geranium), and 5% borage essential oils

50% orange blossom, 40% chamomile, and 10% lilac or violet essential oils

—Stimulating Blends—

75% jasmine, 12.5% sage, and 12.5% citrus essential oils

80% lemon verbena and 20% mint essential oils—either a combination of mints or a single type (such as peppermint, wintergreen, and spearmint)

— Claritive Blends —

60% eucalyptus, 30% chervil, and 10% jasmine essential oils

50% sage, 25% savory, and 25% thyme essential oils

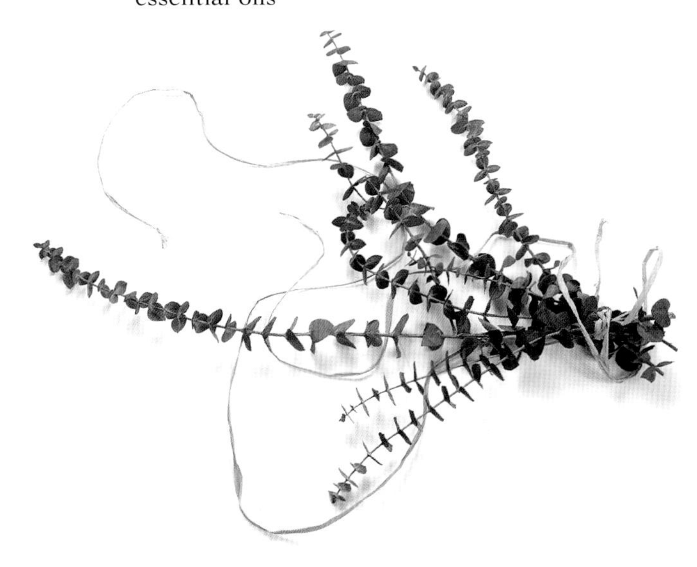

To help you choose which essential oils you would like to try for aromatherapy, the chart on page 48 lists the common responses to specific herbs, flowers, and spices, as determined by current research. You can follow tradition and choose just one essence or create your own aromatherapy blends by combining two or three that produce a similar response. For example, lavender, jasmine, and geranium could be combined successfully because they all evoke relaxation, calm, or restfulness—they are in "mood harmony."

Although the response to any particular smell varies by individual, there are some distinct trends according to sex and age that apply to both aromatherapy and perfumery. If you're planning to make aromatherapy or scented products for gifts and aren't certain about the individual's likes and dislikes, these general trends may be helpful.

Young children and teenagers usually prefer light fruity smells, especially soft, mellow fruits, such as peach, strawberry, and tangerine. (These scents have no known aromatherapy value, however.) When combined with one or more types of fruit, vanilla is almost always a hit with young people and makes an effective tool for teaching them about how fragrances affect their feelings. Be sure to keep all aromatherapy products out of the reach of young children so that there will be no chance of ingestion.

Women generally respond most favorably to scents that are floral, reminiscent of the outdoors, or food-related. Among the three groups, a preference for floral blends predominates. Elderly women usually enjoy pure floral smells that remind them of pleasant events from the past. Rose, gardenia, and jasmine are among their favorites.

Men generally are more responsive to smells that fall into one of these four categories: woody, fresh and clean, citrus, or "baking smells." If you're a woman, you may love the way lavender relaxes you and leaves you in a dreamy state, but the man in your life may prefer to use a jasmine-vanilla blend for the same purpose because it is lighter and fresher smelling and isn't so intensely floral.

Typical Responses to Aromatherapy

SUBSTANCE	RESPONSE
Basil	Stimulating, clarifying
Bayberry	Exciting
Bergamot	Happy
Borage	Calming
Chamomile	Calming, soothing
Cinnamon	Stimulating
Eucalyptus	Creates feeling of balance and sense of well-being
Geranium	Restful
Jasmine	Calming, acts as an antidepressant
Lavender	Relaxing, restful
Lemon	Stimulating
Lemongrass	Fresh, energizing
Lemon Verbena	Stimulating, brings alertness
Lilac	Hypnotic, relaxing
Lily of the Valley	Hypnotic, relaxing
Mixed spice	Inspires contemplation
Orange	Relaxing, calm, adds clarity
Peppermint	Stimulating, creates meditative mood
Rose	Stimulating, strength-ening, slightly hypnotic
Rosemary	Stimulating, uplifting
Sandalwood	Uplifting
Vanilla	Relaxing, sensual
Ylang ylang	Creates balance

Experiment and have fun with your combinations, but as noted earlier, an aroma will work properly only if it has a pleasant or neutral association in the person's memory patterns. Scientists have determined that aromas provide remarkably strong links to our memories, and these can sometimes bring forth powerful emotions. When you use aromatherapy, you're tapping into some of humanity's most basic instincts.

One unintended consequence of your aromatherapy practices may be allergic reactions among your friends or family. Although an essential oil contains no pollen, the strong smell of a plant that causes allergies may be sufficient to start someone sneezing. Go easy until you determine what and how much people can tolerate.

Dispensers

If you're creating aromatherapy products for your personal use, just opening the bottle and smelling your blend or including a few drops in your bath product formulas will probably satisfy your needs. But what if you want to impact more people simultaneously?

Heat is the most common means used for transferring a fragrance or essence into the air for inhalation. Unfortunately, at the same time heat makes a fragrance more accessible, it also damages the essence. For each occasion where you want to dispense aromatherapy into the air with heat, use only what you need and save the rest for later.

There are several types of dispensers specially designed for aromatherapy on the market today. In general, each provides a receptacle for holding a small amount of essential oil and a means for applying gentle heat. Because of the popularity of aromatherapy today, you can find many elaborate, artistic dispensers in specialty and gift stores, and simple lamp rings are available even in some grocery stores.

If you're on a budget or prefer a simpler approach, here are a couple of ways to achieve the same result without much expense.

- Put 6 to 10 drops of your aromatherapy blend in about ½ cup (118 ml) of water in a stainless steel or glass pan on your stove top and turn on the burner to its lowest setting. Soon your kitchen will be filled with the aroma. Remove the pan when the water is almost gone and wash the pan thoroughly after use.

- Apply 1 or 2 drops directly to a light bulb (the bulb must be turned off and cool to the touch); then turn on the light. The heat of the bulb will cause the oil to evaporate, spreading the fragrance into the air. *Caution: Never put an essential oil directly on a hot light bulb; when the liquid contacts the hot glass, it causes the bulb to shatter!* Placing the essential oil on the bulb and warming it slowly to evaporation is much safer and more effective.

If you prefer a less intense fragrance, you can still enjoy the benefits of aromatherapy by using one or more nonheat delivery methods. Add a few drops of your favorite essential oil to some potpourri displayed in an open container in your living room. Scented sachets and fragrance pillows are pleasant accents for your bedroom, and a fragrant tea cozy can add stimulation or relaxation to your afternoon tea.

Chapter Five

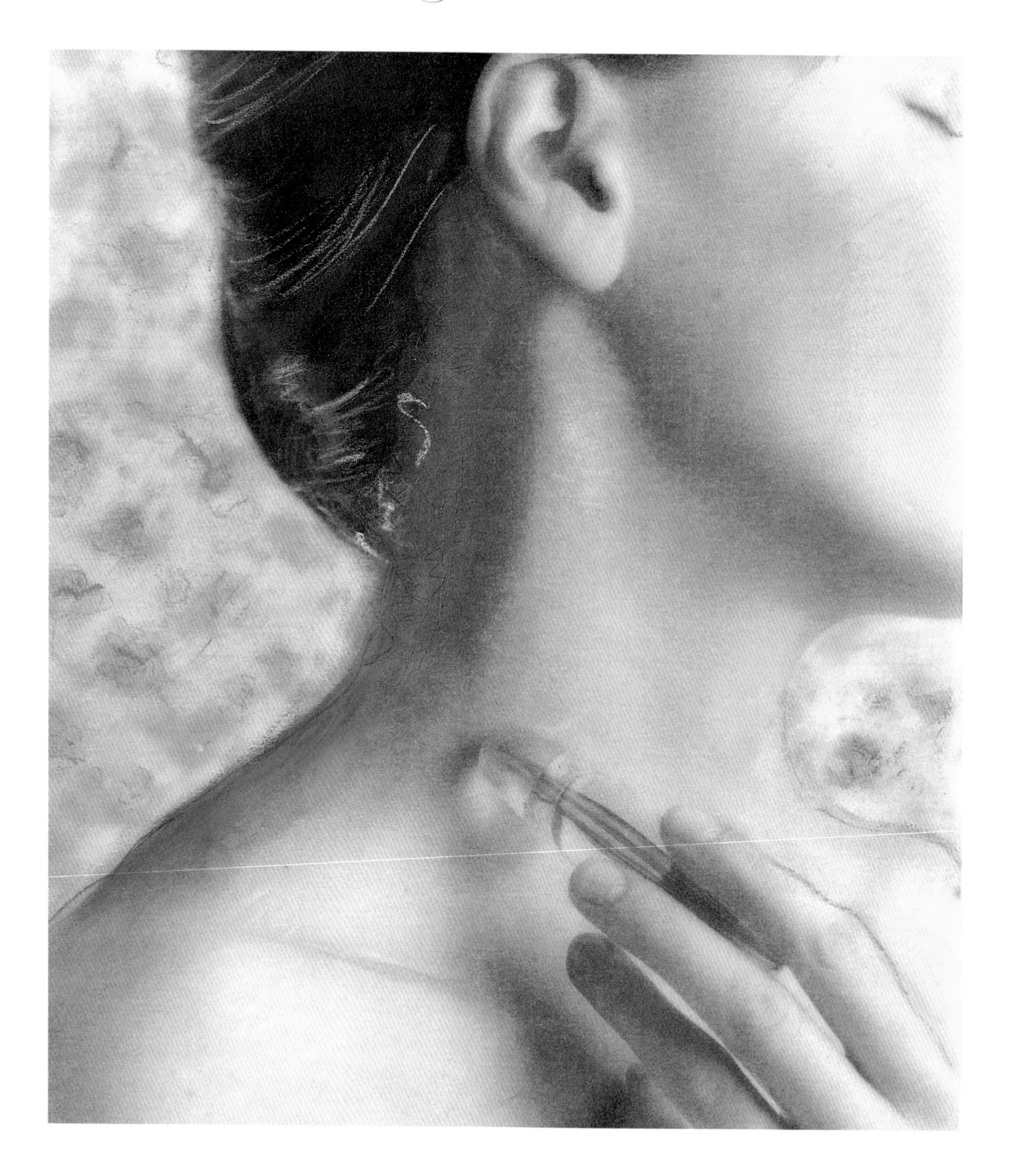

Perfumes & Colognes

PERFUMES AND COLOGNES are traditionally made with alcohol, which is very drying to the skin. Since these products are dotted on pulse points and not applied to the skin in great amounts, the use of alcohol isn't a real problem. Nevertheless, in addition to the more traditional recipes, this chapter includes alcohol-free variations as well.

Understanding & Blending Fragrances

In the world of perfumery, fragrances or essences are divided into general categories. You'll need to recognize these categories before you can decide the main theme for your fragrance. Following the description of each category are examples of some classic blends—a complete guide to fragrances would require an entire book of its own—use these as the first steps in your own experiments in perfumery.

The older, more traditional blends are given in drops, and the newer ones are specified in percentages. When you experiment, keep track of your measurements in parts or percentages—they will be easier to convert to larger quantities.

All of the fragrances noted in these suggested blends are essential oils. When creating blends of two or more essential oils, always use glass mixing containers to avoid spoiling your scents.

FLORAL: Derived from flowers, with either a light or heavy floral smell. A classic example of a floral blend is 15 drops of rose, plus 5 drops of vanilla, and 1 drop of jasmine. Another possible combination is 10 drops of lavender, 4 drops of rose, and 2 drops of vanilla.

OUTDOORS: Fresh and clean, including grasses, balsam, and pine. One of my favorite outdoor blends is to mix 49 percent each of pine and vanilla with 2 percent of jasmine.

HERBAL: Derived from aromatic plants, such as rosemary, sage, and mint, which are often used for medicine or seasoning. Herbal fragrances are very popular, and there are numerous combinations you can try. A classic blend is 60 percent rosemary and 40 percent lavender, but the percentage of rosemary can be increased up to 70 percent, depending on your taste. You can even reverse the two, using 60 percent lavender and 40 percent rosemary, if you prefer a more floral fragrance. For a very bright and fresh herbal scent, combine 14 drops of mint with 6 drops of lemon.

ORIENTAL/EXOTIC: Heady, musky, or deep, heavy scents; examples include patchouli, sandlewood, and musk. A simple formula for an exotic blend is 15 drops of patchouli plus 5 drops of rose. For a more sophisticated scent, mix 20 drops of sandalwood with 4 drops of rose and 2 drops of cedar.

SPICE: Bright, pungent, and zesty, including clove and cinnamon. An appealing blend for both men and women is 10 drops of clove plus 10 drops of cinnamon plus 5 drops of vanilla. Another favorite combination, especially for men, is 80 percent bayberry and 20 percent clove.

CITRUS: Tangy yet fresh; includes orange, lemon, and grapefruit. For a bright, classic citrus blend, combine equal amounts of lemon and lime; then try varying the percentages until you find your favorite. A more subtle alternative is 16 drops of vanilla plus 4 drops of orange.

FRUITY: Any fruit other than citrus, such as peach, kiwi, and strawberry. A wonderful sweet-and-tart mixture is made using equal parts kiwi and strawberry. As with the lemon-lime blend, try changing the percentages a little each time and note the differences in the scent. For a blend that hints of the tropical south, combine 10 drops of peach with 10 drops of mango and 5 drops of vanilla.

Once you've mixed and noted the characteristics of these basics, experiment with the blends by varying the percentages of the components; then try combining fragrances from two or more different categories. For example, you can blend vanilla or jasmine essential oil with almost anything; vanilla tends to soften fragrances, and jasmine is widely used in commercial perfume blends. Usually it's easiest to start by mixing florals with florals and fruits with fruits until you get some experience. Keep in mind that if you change even the slightest proportion, it will dramatically alter your fragrance. Here are some combinations to try:

- Peach with cinnamon or other spice (spicy/fruity blend)
- Raspberry and rose (fruity/floral combination)
- Vanilla and rose (soft floral)
- Lemon, orange, and vanilla (soft citrus blend)
- Lavender and rose (floral mixture)

Now devise your own blends; this can easily consume hours of your day, but it's fun, and people will tell you that you smell wonderful.

To guide your efforts, spend some time in the perfume section of your favorite store. Can you identify some of the individual scents in the blends? Does the blend smell soft, sharp, sweet, or citrus-based? Train your nose to distinguish the differences.

It's very important not to get discouraged while you're experimenting. You may create something that smells terrible at first, but if you change the proportions a tiny amount you may produce something wonderful.

Making Perfume & Cologne

Now that you feel comfortable with blending fragrances and have a few combinations you like, you're ready to make them into perfume or cologne you can wear. If you're starting from scratch with a recipe, mix the essential oils first; then add the diluent to make the perfume or cologne.

The primary diluent for making perfume and cologne is alcohol, and there are two kinds of alcohol that are suitable for home blending. If you want an absolutely pure alcohol (no extraneous ingredients), look for ethyl alcohol, which is 100 percent grain alcohol. Your other choice is to use 90-proof clear vodka. When purchasing vodka for this purpose, select one with the highest alcohol content and least odor possible.

Be sure to use glass containers for mixing and storing your perfumes and colognes.

Traditional Perfume

Combine 1 part pure essential oil (your blend of fragrances) with 4 to 6 parts alcohol. The exact amount of alcohol to use depends on the strength of your essential oil and your personal preference. Shake before using.

Shelf life is 1 to 3 years.

Natural Perfume

If you prefer not to use any alcohol, you can substitute a light oil as your diluent. Combine 1 part pure essential oil with 4 to 6 parts sweet almond oil. Shake before using.

Shelf life is 1 to 3 years.

Note: *Grapeseed oil, apricot kernel oil, or a well-refined safflower oil can all be substituted for sweet almond oil, if desired. Choose an oil with a lengthy shelf life and very little odor of its own.*

Traditional Cologne

Cologne is lighter than perfume, which means only that it is more dilute. Combine 1 part essential oil with 10 to 20 parts alcohol. Again, adjust the exact percentage of alcohol according to your preference and the strength of your essential oil. Shake before using.

Shelf life is 1 to 3 years.

Natural Cologne

Combine 1 part essential oil with 10 to 20 parts sweet almond oil. Shake before using.

Shelf life is 1 to 3 years.

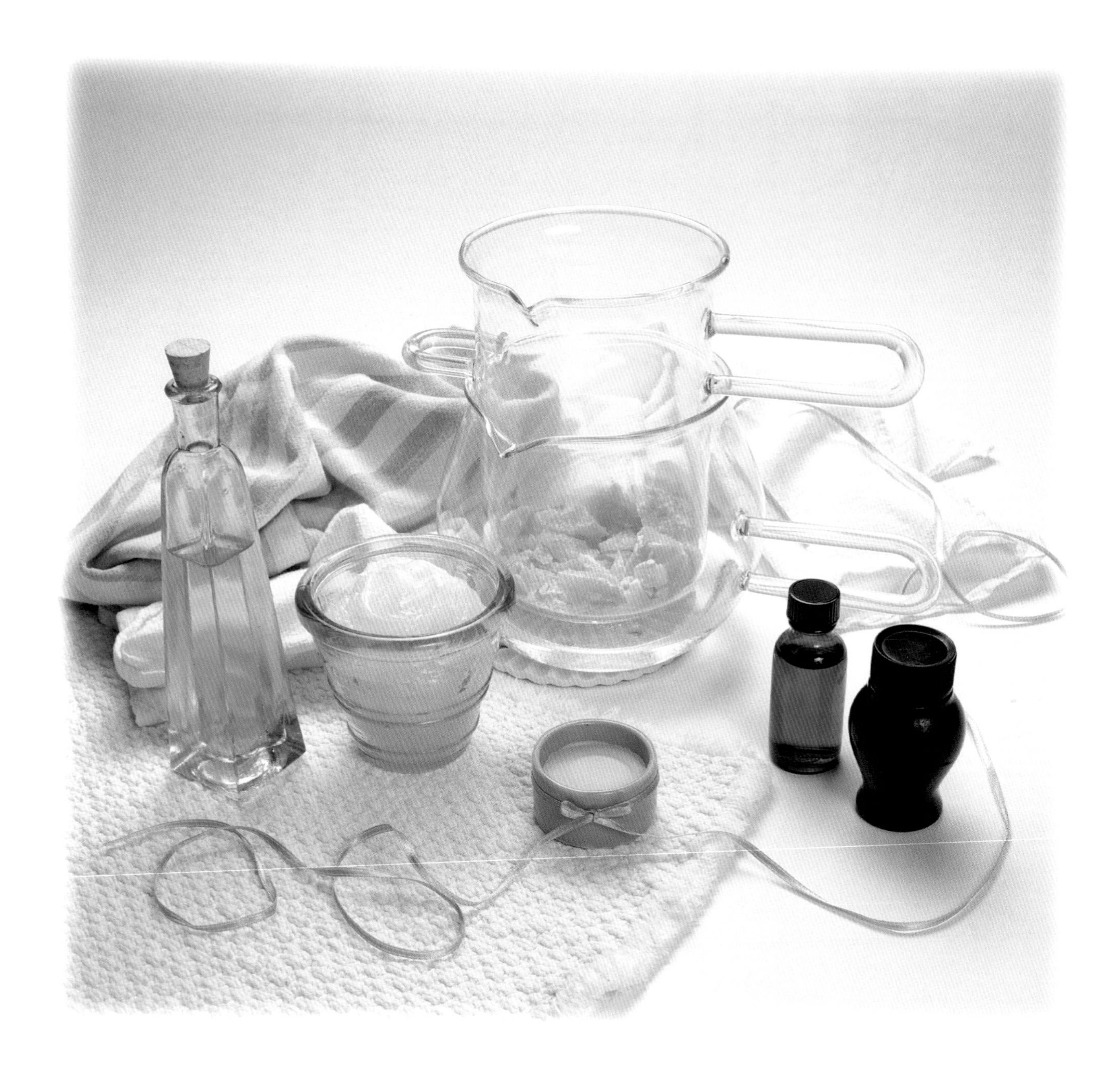

Solid Perfume

This recipe includes petroleum jelly, a heavily refined, man-made ingredient because solid perfume—unlike the other recipes in this book—is not a skin care product. In addition, the consistency of the petroleum jelly helps the perfume glide onto your pulse points more easily.

1 ounce (28 g) beeswax
8 ounces (227 g) petroleum jelly
1 tablespoon (15 ml) essential oil

Melt the beeswax and petroleum jelly in a glass pan or double boiler over low heat. Remove it from the heat and stir in the essential oil. Stir slowly for about one minute; then pour the mixture into decorative containers and let it air-dry until firm.

Shelf life is about 2 years.

Pure Beeswax Solid Perfume

If you prefer not to use any man-made ingredients, here is an all-natural alternative formula.

4 ounces (113 g) beeswax
½ cup (118 ml) sweet almond oil
1 tablespoon (15 ml) essential oil

Using a double boiler or a glass pan over low heat, melt the beeswax together with the oil. Remove it from the heat and add the essential oil, stirring slowly for about one minute. Then pour the mixture into glass jars or decorative tins and allow it to become solid.

Shelf life is 12 to 18 months.

Fragrance Do's & Don'ts

1. Avoid exposure to heat and sunlight, which can damage your product. Keep your fragrance stored in an amber or dark glass container that has a tight-fitting lid.
2. Make it a rule never to test or formulate more than three fragrances at a time to keep your nose from wearing out. Keep some coffee beans handy to sniff so that you can clear your nose between blends.
3. When blending, use a paper strip to test and mix in very small amounts until you have your blend finalized.
4. Keep a detailed log of everything you do. So many variations are possible that if you create something truly wonderful without taking exact notes, you may not be able to replicate it. People have gone mad over less.
5. Make fragrance a way of life and remember that stronger is not always better when you're wearing your new perfumes and colognes.
6. Do not apply fragrance to fine silk or other fabrics that might stain.
7. True essential oils should not be applied directly to the skin without first diluting them with proper diluents.
8. If you blend a scent that smells beautiful in the bottle but not very attractive on you, don't throw it out. Try it on a few friends first; because each person's chemistry is different, it may smell fabulous on someone else.
9. If you're making a product that requires heat during processing, remember that heat damages fragrance. Avoid adding your fragrance during the high-heat phase; add it during the cooling stages.
10. Remember to mix and store your fragrances (essential oils, perfumes, or colognes) only in glass containers; if you use plastic ones, they will collapse.

Chapter Six

The Sensuous Bath

THROUGHOUT HISTORY, BATHING has served many purposes other than just cleansing. It is a favorite method for relaxing and purifying one's thoughts, and in some cultures—notably ancient Roman and contemporary Japanese—it is a form of social interaction.

Moisturizing Bubble Bath

Bubbles are fun and luxurious, and no doubt many a trouble has been floated away during a long soak surrounded by bubbles. Try this lovely combination of bubbles and oils for a delightfully pampering experience.

When you use this recipe in your bath, don't be dismayed if it doesn't produce great quantities of bubbles. Keep in mind that it is a bath oil blend that moisturizes your skin while you soak.

½ cup (118 ml) soap flakes
1 cup (237 ml) boiling water
3 tablespoons (44 ml) sweet almond oil or bath oil blend
A few drops essential oil (optional)

Dissolve the soap flakes in the boiling water, stirring gently. Immediately add the almond oil to the water mixture; then add the essential oil, if desired. Stir constantly while you add 4 to 5 tablespoons of the bubble solution to the running water in your bath. The rest can be used later, but you will need to stir it again, since oil and water don't mix easily.

Shelf life is about 1 month at room temperature.

Hint: *Make your own soap flakes by saving all the little bits and pieces of used soap until you have accumulated a cup or two (237 to 473 ml). Your own handmade natural glycerine soap is ideal for this (see page 39). Using a blender, food processor, or hand grater, grind the pieces to a powder. This is an excellent way to recycle those otherwise wasted soap bits.*

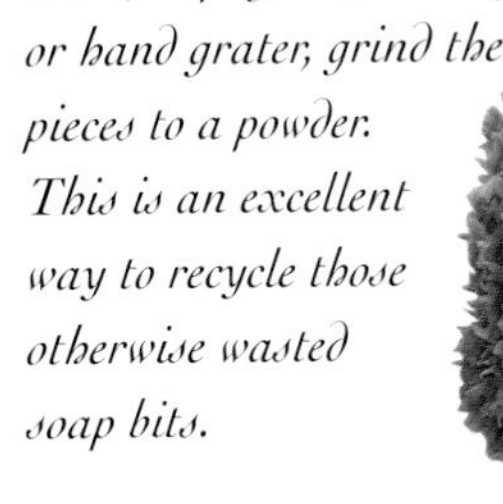

"Big Bubble" Bubble Bath

Large bubbles are more difficult to produce, and most home recipes will make small but adequate bubbles. Glycerine is used in this formula to "stretch" the bubbles and make them bigger.

1 cup (237 ml) soap flakes
1 cup (237 ml) water
2 to 3 tablespoons (30 to 44 ml) glycerine
A few drops essential oil, if desired

Mix all of the ingredients together and store your bubble bath in a beautiful glass bottle or container. Add 1 to 2 tablespoons (15 to 30 ml) under rapidly running water to create large, generous bubbles. This mixture can also be used as a shower gel; shake it before using. Store at room temperature.

Shelf life is 2 to 3 months if stirred occasionally.

Note: *Some people are very sensitive to bubble baths. If you think that you or the recipient of your gift is sensitive to sitting in bubbles, make and package this as a "shower gel" or make the stress-relieving salts or milk bath formulas instead.*

Hint: *Glycerine helps make bubbles bigger by giving the soap base more elasticity. For giant blow bubbles for kids of all ages (not for use on the body), add a small amount of glycerine to normal liquid dish detergent. Make a giant wand from a coat hanger or use the plastic circles cut from a 6-pack holder of canned drinks.*

Bathing Rituals in the Land of Milk & Honey

Milk and honey are time-honored ingredients that have been used for centuries, either as main ingredients or as formula additions. Milk is noted for its soothing qualities (ancient peoples used to think it would also make their skin "milky white") and honey for its skin softening and protective qualities.

Ancient-Formula Milk Bath

It is said that Cleopatra used a formula similar to this one to soften her skin. Each ingredient is known for its soothing and smoothing qualities.

1 cup (237 ml) powdered milk
2 tablespoons (30 ml) almond meal
2 tablespoons (30 ml) barley or oat flour
A few drops of your favorite essential oil (optional)

Mix all of the ingredients in a bowl. This mixture can be sprinkled directly into the bath, using 2 tablespoons (30 ml) per bath. An interesting variation is to substitute rolled barley or oat flakes for the flour, then fill empty tea bags with the mixture (remember to seal the tops). Soak in your aromatic milk bath for approximately 20 minutes. When you're finished, wrap yourself in a warm towel and enjoy the cleanliness and softness this bath imparts to your skin.

Store at room temperature in pretty tins with lids or in old milk bottles.

Shelf life is 2 to 3 months.

21st Century Milk Bath

In this current-day natural formula, milk, baking soda, and cornstarch are used for soothing and smoothing, and the epsom salts soften both water and skin. (You can substitute sea salt for more of a toning effect, if you desire.)

1 cup (237 ml) powdered milk
½ cup (118 ml) epsom salts
1 tablespoon (15 ml) baking soda
1 teaspoon (5 ml) cornstarch
A few drops of your favorite essential oil (optional)

Mix all of the ingredients in a bowl. This blend, like Cleopatra's version, can be used directly in the bath or packaged into tea bags, and it has the same storage requirements and shelf life.

Hint: *If you can't find empty tea bags (usually available at health food stores), then tie the mixture securely in cheesecloth or a washcloth and secure it with a pretty ribbon.*

Hint: *Add simple soap flakes to your milk bath to make a foaming bath version. (You can buy soap flakes at drug stores or make your own—**see the recipe for bubble bath on page 59.**) For a beautiful effect, add dried flower petals or herbs to either formula.*

No Weight-Gain Bath Pudding

Since you will be soaking in it and not spooning it into your mouth, this pudding is guaranteed not to add inches around your waist. This recipe should be made up fresh each time you plan to use it, so pour the entire mixture into your bath. It's a combination milk bath/bath oil formula that smoothes, conditions, moisturizes, and pampers your skin.

2 egg yolks
¼ cup (59 ml) sweet almond oil
2 cups (473 ml) distilled water, warm
1 teaspoon (5 ml) milk
A few drops essential oil (optional)

Beat the egg yolks until frothy. Slowly add the milk and sweet almond oil to the beaten eggs and mix thoroughly. Gradually add the warm water, stirring until everything is well blended. Add a few drops of fragrance last, stir, and immediately add the pudding to your bath. Then settle in for a wonderful treat.

Use immediately; bath puddings have no shelf life.

Variations: *For an old-fashioned buttermilk bath, substitute 1 cup (237 ml) buttermilk for the egg yolks. If you want the ultimate treat, dissolve 1 tablespoon (15 ml) of honey in the distilled water prior to adding it to your mixture. (Remember, the honey will dissolve more quickly in warmer water.)*

Hint: *When you separate the eggs for this formula, beat the whites until they're firm and apply them not to a pie top, but to your face. Enjoy an egg white facial while you soak. (Egg whites firm and tighten the skin and wash off easily.)*

Stress-Relieving Bath Salts

The concept of soaking in salt and mineral waters connects us with the earth's mountains and seas. Minerals known to relax the body are combined with sea salt, which softens the water and tones the skin. This relaxing formula combines Mother Nature's salts with hydrating oils to moisturize your skin.

1 cup (237 ml) epsom salts
¼ cup (59 ml) concentrated sea salt
1 teaspoon (5 ml) sweet almond oil or apricot kernel oil
A few drops essential oil of lavender or other relaxing scent (optional)

Mix the salts together; then stir in the almond or apricot kernel oil, a few drops at a time, until it's evenly distributed. Add the essential oil if desired. Use ¼ cup (59 ml) of the salts per bath, adding them under running water. Store in a pretty container with a lid and a scoop.

Shelf life is 2 to 3 months.

Floral Bath Vinegar

For years herbalists have enjoyed the toning and aromatic effect of herbal bath vinegars. This recipe calls for roses, which are among the most aromatic flowers.

1 cup (237 ml) white vinegar
1 cup (237 ml) distilled water
2 cups (473 ml) fresh rose petals

Mix the vinegar and water; then add the rose petals and place the mixture in a glass bottle with a screw top. Shake the bottle and store it in a cool, dark place for one month (shaking it every week or so). After a month, shake one last time; then strain out all the flower remains. Now place your floral bath vinegar into a decorative glass bottle.

Shelf life is 2 to 6 months if kept sealed.

Variations: *For the rose petals, you can substitute 3/4 cup (177 ml) of lavender flowers, 1 cup (237 ml) of cut lemon and orange peels, or 1/2 cup (118 ml) of dried fresh herbs. Chamomile flowers are calming, rosemary clarifies the skin, mint leaves are stimulating, and lemongrass or lemon balm are good toners.*

Herbal Bath Tea

A less pungent alternative to a bath vinegar is a gentle herbal bath tea (infusion). When making an infusion for the bath, recall that it will be heavily diluted by the bath water and should be steeped until it's fairly strong in fragrance.

Tie ½ cup (118 ml) of your favorite herbs into a piece of cheesecloth or a spare nylon knee-high stocking. Bring 2 cups (473 ml) of water to a boil in a saucepan on the stove; then remove the water from the heat. Drop in the pouch of herbs, cover the pan, and let it steep for 15 to 20 minutes. Remove the cover, taking care not to burn yourself, squeeze any remaining liquid from the packet of herbs into the infusion, and discard the spent herbs. Once your infusion has finished cooling, you can bottle it in decorative glass containers.

Shelf life is 2 to 6 months if kept sealed and if dried (not fresh) herbs were used.

Hint: *For added beauty in your herbal vinegars or infusions, you can suspend a sprig of the dried herb in the solution. This can be done successfully with most herbs and citrus. However, with the exception of lavender sprigs, flowers are not a good idea because they mold easily due to their high water content.*

After-Bath Dusting Powders

Bath and dusting powders first came into vogue in the Elizabethan era, when fragrant powder was rubbed onto the outside surfaces of gloves to perfume them. Whenever acquaintances met and shook hands, a gentle fragrance was released between them. (Today this technique could come in handy for quick touch-ups to an oily face.)

Natural Body Powder

In more recent years, talcum powder has been used directly on the skin as an effective absorbent, to help deodorize, and for imparting a silky touch. We realize today, though, that talc (talcum) is not healthy to breathe because it often contains traces of asbestos. This is a lovely talc-free powder.

½ cup (118 ml) cornstarch
4 tablespoons (59 ml) rice flour
6 drops of your favorite essential oil

Stir the cornstarch and rice flour together. Using a hand sifter, sift ¼ cup (59 ml) of the mixed powder with 2 to 3 drops of essential oil. Repeat with the remaining powder and essential oil. Then sift the whole powder mixture a second time. Put it in an herb shaker or a pretty box with a powder puff. This powder is well suited for infants, and it works nicely in your shoes on hot summer days or in sneakers anytime to help absorb excess perspiration.

Shelf life is 1 to 2 years at room temperature.

Hint: *If the powder loses some of its fragrance over time, sift it again with a few drops of essential oil. You can also add crushed dried flower petals or herbs to give your powder more fragrance and a decorative appearance.*

Extra-Fine Body Powder

This is a finer powder than the previous recipe—perfect for women who want just a light touch. You can find lots of pretty containers for your powder: glass cheese shakers, antique powder containers, and decorative tins are just a few possibilities. Add a few dried flower petals or herbs to your powder for visual effect and additional scenting.

¼ cup (59 ml) arrowroot powder
¼ cup (59 ml) cornstarch
3 tablespoons (44 ml) rice flour
6 drops essential oil

Follow the same steps as described at left.

Shelf life is 2 years or longer if kept covered and stored in a dry place.

Hint: *Low-cost powder puffs are available at drug stores. If you have access to sheep and do wool working, clean and card the wool, roll it into a puff ball, and tie a knot at the top for a handle. This makes a very special handmade gift.*

Spa-Style Herbal Body Wrap

Herbal wraps act like an herbal sauna and are used to detoxify the body, causing it to sweat out impurities. To make the large quantity of liquid required, you can convert your sink into an herbal infusion pot or use a very large soup kettle on top of your stove. The herbs included here are the same as you would use for making regular bath infusions.

6 cups (1.4 l) very hot water
½ cup (118 ml) dried chamomile
½ cup (118 ml) dried rosemary
¼ cup (59 ml) fresh mint or peppermint leaves
Cheesecloth or wide cotton gauze
Linen or loose-weave cotton, cut into long strips 4 to 6 inches (10 to 15 cm) wide
Roll of plastic wrap

Bring the water to a boil and remove it from the heat (or pour it into the stoppered sink). Wrap the herbs in the cheesecloth or gauze, making a large bundle, and push them down into the water with a large spoon or other utensil. Cover the infusion to contain the steam and allow it to sit for 15 to 20 minutes. Uncover the herbal infusion, stir it once or twice, and add the long strips of linen or cotton. When the fabric is saturated, wring it out and wrap the strips around your body (like a mummy) except your head, lower arms, and feet. The infusion and fabric should be hot, but not so hot they will burn you! When you're done, repeat the wrapping process by covering the fabric with the plastic wrap.

Make a cold compress for your head and pour yourself a cool glass of water to sip while you rest. Now lie down, relax, and let the herbs do their work. Leave the wrapping on your body for no more than 15 to 20 minutes; then unwrap and cool your body for another 15 minutes before showering.

Hint: *Just as in a sauna, an herbal wrap surrounds your body with a higher than normal level of heat. If you have high blood pressure, heart problems, or other medical conditions, such high-heat situations may be harmful. Be sure to consult your doctor if you have health problems and want to do this procedure. For the average healthy person, this is a great way to "sweat out" the impurities accumulated in your skin.*

Chapter Seven

Body Creams & Lotions

EVERY DAY, INDOORS AND OUT, your skin is living in a war zone. It's under constant assault by the drying effects of central air conditioning and heating, pollutants in the air, seasonal temperature changes, the sun, wind, and more. Gently cleansing your skin and creating a moisture barrier to help protect it are your best defense methods. Here are some natural creams and lotions to help you create that protective barrier.

Types of Creams & Lotions

The classic definition for each major type of cream and lotion is given below. The differences among them are less distinct today than they once were because of new manufacturing techniques and the extensive range of new polymer-based emollients and hydro-emollients. For example, the viscosity levels of certain products are now somewhat misleading. (Viscosity refers to the thickness of a product and usually relates to the amount of oil added and the moisturizing capacity of the preparation.) Because of today's technology, some seemingly heavy creams can have high levels of alcohol and other ingredients that dry out the skin, and some lighter-feeling preparations may be packed with emollients.

HEAVY CREAMS: These are most often cleansing creams and/or make-up removers. Traditionally they contain the highest oil content and tend to sit on the skin the longest, giving it a wet appearance. For this reason, even though they are excellent moisturizers and softening agents, heavy creams have been deemed unsuitable for normal facial applications; it is almost impossible to apply make-up over them. In addition to their skin-softening abilities, heavy creams also loosen dirt and old make-up, making them ideal for use as cleansers. Cold creams and some make-up removers, night creams, and eye creams are considered heavy creams. These products are usually so thick that they are presented in jars.

CREAM: A good face, hand, or body cream is loaded with emollients, and unlike a heavy cream, it is absorbed into the skin within an acceptable period of time (there is no formal agreement on how much time is *acceptable*). A cream has less water than a standard lotion and is usually so thick that it is packaged in a jar. In a bottle, it would not pour or dispense well.

LOTION: This is a lighter version of a cream, and it usually has a higher water or polymer content to make it easy to dispense from a bottle. Lotions are used for applying moisture to the face, hands, and body when quick absorption is desired.

SPLASH, TONIC, TINCTURE: Having either a water or alcohol base, these are all very low in viscosity (thin consistency). They can be applied by splashing them onto the face and body with your hands or by dabbing them onto the skin with a cotton ball. A splash, tonic, or tincture can also be used in a spa-style body wrap.

RUB: As its name suggests, this is a preparation that is meant to be rubbed onto the body. It can have the consistency of a light lotion or a splash/tonic. A rub is usually a healing preparation of some kind and is often formulated for soothing sore muscles. In commercial preparations, the terms *rub* and *salve* are sometimes used interchangeably.

SALVE: The main purpose of a salve is to provide relief from discomfort or to heal minor skin problems. It usually has a waxy consistency so that it glides onto the skin. The wax acts as a carrier for the healing ingredients, keeps the affected area supple, and creates a protective barrier to guard against water penetration.

In this chapter there are more recipes for creams than for lotions because creams are easier to create at home. Natural lotions require a higher water content, which makes the emulsifying process more difficult and increases the risk of bacteria growth. (A less natural approach is to use mineral oil, a petroleum distillate.) Creams can be made with natural oils or emollients that will give you a significantly longer shelf life.

Soothing Cucumber & Aloe Cream

The juice from the fleshy leaves of the aloe plant has long been recognized for its soothing properties, and cucumber is noted for refreshing and toning the skin. The oils are included for their moisturizing and smoothing properties; beeswax and honey provide heavy smoothing and softening. This cream is great for both hands and body.

2 tablespoons (30 ml) aloe vera gel
1 cucumber, peeled
2 tablespoons (30 ml) apricot kernel oil
1 tablespoon (15 ml) coconut oil
4 tablespoons (59 ml) melted beeswax
1 teaspoon (5 ml) honey

Cut the cucumber into several pieces and purée it in your food processor or blender. Strain out the liquid and add the aloe vera gel to the strained cucumber and mix well. Melt the beeswax with the honey in a double boiler on the stove. When the wax is melted, slowly add both oils while stirring; then slowly stir in the cucumber and aloe combination. Remove the mixture from the heat and cover it; then stir slowly about every 4 to 5 minutes until it's cool.

When the mixture is completely cool, store it in the refrigerator.

Shelf life, when kept refrigerated, is 60 to 90 days.

Variations: *For a skin-bleaching cream, follow the same formula as above, but add 2 tablespoons (30 ml) of freshly squeezed lemon juice while cooling and stirring.*

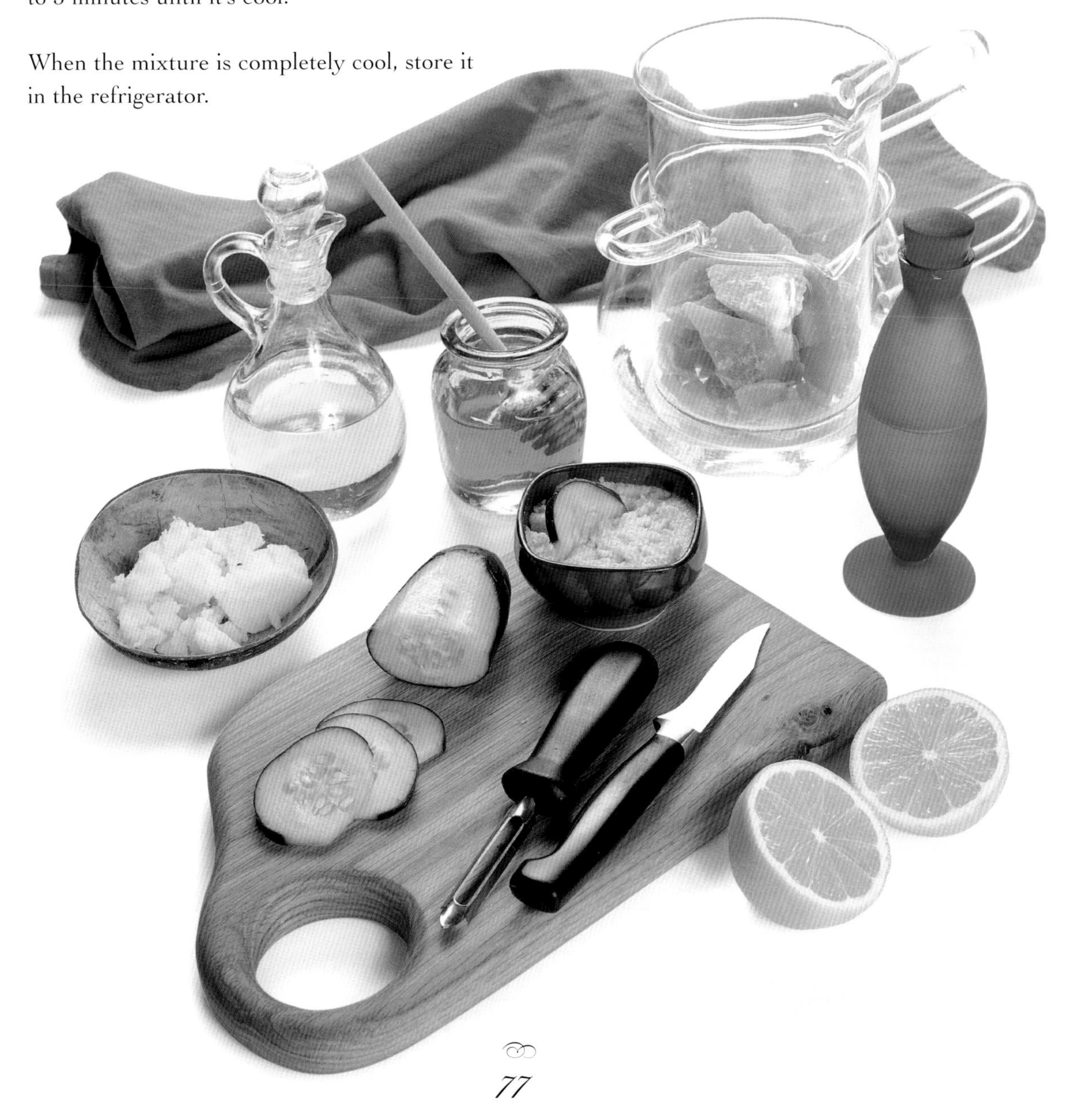

Creamy Rose Moisturizer

Benzoin is a natural resin that takes the place of artificial and petroleum-based products in many commercial preparations. In this recipe it serves as a mild preservative and helps emulsify the other ingredients.

1 cup (237 ml) sweet almond oil
½ cup (118 ml) beeswax
¾ cup (177 ml), total, rosewater and glycerine blended together
¼ teaspoon (1.25 ml) tincture of benzoin
A few drops essential oil, if you wish

In a glass pan on low heat, melt the oil and wax together, stirring often. Remove from the heat and slowly add the blend of rosewater and glycerine, rapidly stirring by hand until the mixture is evenly textured and cool. Add the tincture of benzoin, followed by a few drops of essential oil, if desired. (Because of the ingredients, the essential oils of rose or lavender work best for scenting this wonderful creamy lotion.) No special storage is required.

Shelf life is 6 to 9 months.

Moisturizing Sports Rub

Follow the instructions for the creamy rose moisturizer, but add 4 to 5 drops each of peppermint, wintergreen, camphor, eucalyptus, and menthol essential oils. *Caution: Make certain to keep these essential oils from coming in contact with any mucus membranes (eyes, nose, etc.).* Although the warming sensation created by this rub feels wonderful on certain parts of the body, such as the calves of your legs or your lower back, it becomes a burning, irritating experience if it gets into your eyes or comes in contact with sensitive body parts. Remember to wash your hands thoroughly when using products (homemade or commercial) with these ingredients.

Rosewater & Glycerine Skin Softener

This old-fashioned Victorian recipe is simple yet effective, and it's especially beautiful when displayed in an elegant antique glass bottle. (You may wish to add 1 to 2 drops of food color for effect.) Rosewater and glycerine have long been used for softening skin, and the words alone bring to mind a time when life was far simpler.

½ cup (118 ml) rosewater
½ cup (118 ml) glycerine

Mix together thoroughly and place in a glass bottle. Store at room temperature.

Shelf life is 3 to 4 months, sometimes more.

Hint: *The reason for the short shelf life is, alas, bacteria. Although I prefer not to use alcohol on the skin, the addition of 1 teaspoon (5 ml) of 90-proof vodka will help retard bacteria growth and give this recipe a longer shelf life. (It usually adds another 3 to 4 months.)*

Deep, Intensive Moisturizing Cream

This formula is ideal for elbows, knees, and other trouble spots that seem especially dry. As you rub it onto your body, your hands will enjoy the treatment as well. It can also work as a massage cream.

¼ cup (59 ml) cocoa butter
¼ cup (59 ml) sweet almond oil
¼ cup (59 ml) olive oil
5 to 6 drops essential oil (if desired)

Combine all the ingredients in a double boiler and warm them until the cocoa butter is completely melted. Remove from the heat and stir until the mixture is cool and starts to gel; then allow it to rest as it finishes setting up. When it's cool, the mixture should be creamy. If it separates, then use a hand mixer on low speed and beat until it's smooth. Store in a decorative container. If it's kept in very cool temperatures, this cream will need to be reheated before using.

Shelf life is 3 to 4 months.

Massage Cream

Try this deep massage product as an alternative to massage oil. You can add small amounts of warming ingredients (such as wintergreen, camphor, or peppermint) or other herbs to enhance the formula. Don't hesitate to be creative.

½ cup (118 ml) beeswax
¼ cup (59 ml) sweet almond oil
24 tablespoons (355 ml) distilled water
12 to 16 drops essential oil (if desired)

Melt the wax in a glass container on low heat, stirring constantly. Using a whisk, slowly stir in the almond oil, then the water, a few drops at a time. Blend thoroughly and remove from the heat. Add the essential oil, stir, and pour into containers. This rub is quite solid when it cools; warm it just enough to soften it before using.

Shelf life is 1 year or longer.

Olive "Creamy" Rub

Use this rub to create a protective barrier on newly moisturized skin. This formula can also be used as a massage rub.

½ teaspoon (2.5 ml) borax
2 teaspoons (10 ml) boiling water
2 tablespoons (30 ml) beeswax
4 tablespoons (59 ml) olive oil

Stir the borax into the boiling water until it is dissolved. Melt the beeswax and oil in a double boiler on low heat, stirring it until it is smooth; then add the water-borax mixture. Remove from the heat and stir, continuing to stir until the mixture is too thick to continue. Place it in a decorative tin or glass container and cover.

Shelf life is 6 to 12 months.

Variations: *To make a bug-repellent cream for gardeners or those who love to fish, follow the same formula as above but add citronella essential oil. Substitute camphor or menthol for the citronella to make a sports rub that warms and cools sore muscles.*

Chapter Eight

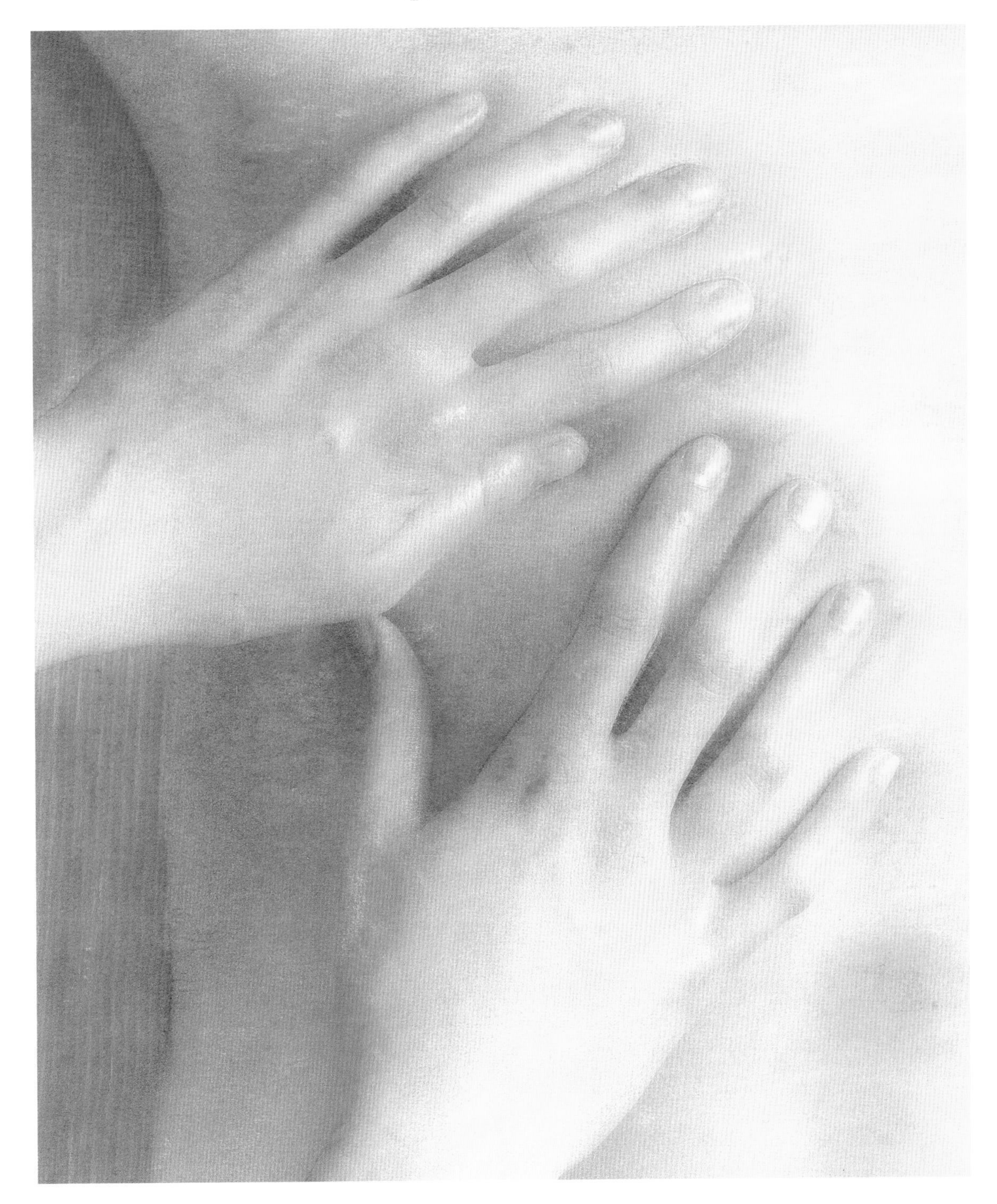

Massage & Body Oils

In the rituals of bathing and relaxation, natural vegetable oils have long played a part. The phrase "anointing in oil" dates back at least to biblical days, when blends of natural oils were used for bathing, relaxing muscles (therapeutic massage), and soothing feet. In addition to these uses, modern applications include under-eye wrinkle reducers, hand and nail treatments, hair conditioners, and suntanning aids, among others.

Natural Oils for the Body

Natural oils used for the body can vary by weight, moisturizing qualities, and suggested uses. Not to be confused with essential oils or fragrance oils, these oils often have no smell (a preferred quality in formulas where you wish to add an essential oil to provide a fragrance) or a slightly pungent fatty smell. Oils used for the body are best if they're unrefined and derived by cold pressing. Heat processing can damage the natural vitamin content of some oils, and refined oils may have been extracted by chemical solvents.

Any natural oil can be damaged by extreme temperatures, either cold or hot. Store your oils in a cool (45 to 65°F/7 to 18°C) area for best results. At cooler temperatures natural oils tend to become cloudy, and you may think they have spoiled. Before you throw out a cloudy oil, warm it to room temperature. If it becomes clear again and smells fine, then it's still good. Most natural oils have a shelf life averaging from nine months to one year.

Some of the most frequently recommended types of oil for use on the body are listed here. Most body oil formulas are blends of two or more of these oils. Experiment with the proportions yourself or use the recipes provided in this chapter.

—Base Oils—

Base oils are used as the main ingredients in massage and bath oil combinations or for blending with or diluting essential oils.

Apricot kernel oil: Practically odorless; very light oil; used in body and hair applications; available in health food stores

Grapeseed oil: Has slight odor; good for body or hair; sold in health food stores

Macadamia nut oil: Slight odor; used for body; found in health food and grocery stores

OLIVE OIL: Distinct and pungent aroma; heavier than other base oils and more commonly used in tropical locations for hair or after-sun applications; sold in health food and grocery stores

SAFFLOWER OIL: Another heavier oil, with a noticeable, slightly nutty scent; used in massage oil blends; available in health food and grocery stores

SUNFLOWER OIL: Works well as a bath or body oil; light in density, fragrance, and color; sold in health food stores and some grocery stores

SWEET ALMOND OIL: Very light, nearly odorless oil; used for body or hair; found in health food stores and pharmacies

Hint: *Sweet almond oil, apricot kernel oil, and grapeseed oil are my favorites to use alone or as base oils in blended formulas because of their nourishing qualities. They all have excellent moisturizing properties, are easily absorbed by the skin, and are recommended for all skin types.*

—Suggested Additive Oils—

Some oils are better suited to be used in smaller quantities than base oils. They're valued for their moisturizing and "glide" properties.

BORAGE SEED OIL: Contains gamma-lanolin acid, which has been recognized as an aid in regulating the body's metabolism; often sold as a mixture with evening primrose oil at health food stores

CALENDULA OIL: Soothes sensitive and irritated skin; often used in children's products; available in health food stores

COCONUT OIL: Heavy oil commonly used in tropical locations for conditioning hair or soothing sun-drenched skin; available at health food stores and pharmacies. Coconut oil becomes solid in cool weather, and you may need to heat it before adding it to your mixture. If, when added as a liquid, coconut oil does not constitute more than 30 percent of your blend, then it will usually remain liquid when cool. Because it is such a heavy emollient, coconut oil is used undiluted by some therapists for massage.

EVENING PRIMROSE OIL: Also contains high levels of gamma-lanolin acid (see borage seed oil); found in health food stores

HAZELNUT OIL: Good for dry, damaged skin; has a slightly nutty fragrance; available at health food stores and some grocery stores

JOJOBA OIL: Makes an excellent base for perfumes, facials, and skin oils; has good anti-inflammatory properties but can be irritating to some people; longer than normal shelf life; odorless; sold at health food stores

ST. JOHN'S-WORT OIL: Contains vitamin E and is great in blends for sunburn and ordinary burns; may be less readily available but can be found in some health food stores

WHEAT GERM OIL: High in lecithin and vitamins (including vitamin E); helps preserve oil blends (if you use at least 15 to 20 percent); good for dry skin; strong odor; available in health food stores

Look for other specialty oils available at your local health food store and inquire about their usage; you may find others listening in with you.

Adding a Fragrance

When using essential oil(s) to provide fragrance to your blends, the standard rule is to add an amount that represents between ½ percent and 2 percent of your formula, depending on how strong you want the fragrance to be. If you're doing a large batch, you can calculate the correct amount by multiplying the total number of ounces of your finished product by the percentage you want to add. If you are blending a small amount, such as 1 cup (237 ml) or less, simply add approximately 5 to 10 drops of essential oil.

Before adding any essential oils to your massage or body oil blends, make sure they are safe for use on the skin.

Packaging Massage & Body Oils

Colored bottles help protect your natural oils from vitamin loss due to exposure to light, but you may wish to use clear bottles for your floral oil blends so that you can enjoy the beauty of the flowers held in suspension. The bottles can be as simple or ornate as you wish, but all should be thoroughly cleaned before they're used. If you're using a bottle with a cork stopper and want to ship it to a friend, dip the top several times in hot paraffin, letting the wax cool between dips. In addition to its quaint, old-fashioned look, a wax seal keeps the bottle from leaking during shipment.

Hint: *For a special touch when sealing with wax, press a few flowers, piece of ribbon, or other decorative items into the soft wax immediately after the last dip. Don't forget to label your creation with the name of your special blend and its suggested use.*

Body Oil

This combination is ideal for skin that has been exposed to too much sun or for extra-dry areas, such as heels and elbows.

1 cup (237 ml) sweet almond oil
½ cup (118 ml) jojoba or hazelnut oil (or combination of the two)
2 tablespoons (30 ml) apricot kernel oil
Essential oil (optional)

Combine the oils in a sealed bottle and gently turn it several times to mix.

Pregnancy Oil

This formula helps keep skin moisturized and supple to help guard against stretch marks.

½ cup (118 ml) sweet almond oil
½ cup (118 ml) apricot kernel oil
½ cup (118 ml) coconut oil
½ cup (118 ml) calendula oil
Essential oil (optional)

Combine all the oils in a pretty bottle with a secure lid or stopper. Turn the bottle several times to mix thoroughly. Rub the oil onto the expanding areas twice daily and after bathing. Apply it in small amounts and rub until the oil is completely absorbed—your hands will benefit too!

Deep-Tissue Massage Oil

Massage therapists favor blends similar to this for their ability to stay active on the skin long enough to allow them to work with the muscle tissue. Over time the oil will absorb into the skin and moisturize it.

1 cup (237 ml) sweet almond oil
¾ cup (177 ml) apricot kernel oil
⅔ cup (158 ml) grapeseed oil
2 tablespoons (30 ml) safflower oil for added weight
Essential oil (optional)

Combine all the ingredients in a bottle and seal it tightly. Turn the bottle a dozen or so times until the oils are well mixed. Apply at room temperature or slightly warmed.

Sports Relief Oil

In this formula, the essential oils are active ingredients and are necessary for relief of sore muscles. (This oil is ideal for an athlete's weary feet and legs.) Don't try to add more fragrance to the combination—it is quite commanding on its own.

1 cup (237 ml) sweet almond oil
½ cup (118 ml) sunflower oil
3 tablespoons (44 ml) calendula oil
8 drops each of peppermint, wintergreen, and eucalyptus essential oils

Combine the oils in a stoppered bottled, turning it over several times to mix them thoroughly. Use after a workout by massaging it into any stiff or sore muscles.

Floral Oil Suspension

This blend can be used as a bath or body oil. Because of the beautiful effect created by the flowers, this oil is a popular gift idea.

1 cup (237 ml) sunflower oil
or apricot kernel oil
½ cup (118 ml) sweet almond oil
½ cup (118 ml) wheat germ oil
Essential oil (optional)

Choose the flowers you wish to add (petals, flower heads, or complete flowers on stems, depending on look you want). Use only flowers that are safe for use on the body, such as rose, larkspur, globe amaranth, baby's breath, and strawflower. You can also use small leaves or pods that are nontoxic.

All flowers must be completely dried (not fresh) and free of any insects, debris, or additives. If you don't want to dry your own flowers, you can buy them from specialty growers, order them from a local

florist, or find them at harvest festivals. Do not use flowers from commercial potpourri, since most are treated with a fixative that will cloud the oil. In addition, most commercial fixatives are unsafe for use on the body.

Arrange a few flowers in your bottle, remembering that some of them will float and move around. Single petals are likely to float until they become saturated. Flowers on stems can be placed more precisely because they are more likely to stay put. A combination of stems and petals or flower heads provides some movement and some stable artistic arrangement. Don't overfill the bottle—a pleasing color combination of a few blossoms showcased in the oil will look better than many pressed together.

Now fill the bottle with the blended oils, adding the fragrance last, if desired. The flowers themselves won't impart any fragrance to your oil because they are dried; however, the oil will preserve the flowers and make them look beautiful.

Gently shake a few times before using.

You can substitute herbs for the flowers, but dried herbs will sometimes add fragrance to your oil blend. Examples of herbs to substitute include rosemary, chamomile, mint, wintergreen, and sage.

Chapter Nine

Pampering Your Face

FOR YEARS WOMEN HAVE FOUGHT the signs of aging on their most visible assets—their faces. More recently, men have also learned the benefits of taking greater care. Here are some natural formulas for moisturizing, firming, cleansing, and toning your facial skin. Always remember to be extra gentle with the skin around your eyes; don't apply facial masks or astringents to this delicate area.

Natural Homemade Cold Cream

In the 16th century Galen invented cold cream, and amazingly enough, the formulas of today do not differ much from his original. The major difference is the substitution of mineral oil (a petroleum product) for natural beeswax. Every ingredient in traditional cold cream has a specific purpose: the wax and oil are used to soften dirt and make-up for easy removal from the skin; water makes it smooth enough to use; the borax and the cooking procedure help bind the formula together.

Cold cream got its name not from anything relating to temperature but from the fact that it was originally made with coleseed oil. The original name, *cole cream,* gradually became cold cream simply through repeated mispronunciation. Beginning in the Victorian period, sweet almond oil has been substituted for coleseed oil because it is easier to obtain.

This is a Victorian recipe that I have adapted to give it a bit more nutritional value for the skin.

½ cup (118 ml) sweet almond oil
5 tablespoons (74 ml) water or rosewater
2 tablespoons (30 ml) melted beeswax
¼ teaspoon (1.25 ml) borax

Put the wax and oil into a double boiler and stir over a low heat until they are melted together with an even consistency. In a separate saucepan on a medium setting, heat the water or rosewater and stir in the borax until it's completely dissolved; then remove it from heat. While beating vigorously (using a spoon or electric mixer), add the water-borax mixture a little at a time to the blend of wax and oil until all the water has been added. Remove the mixture from the heat and continue to beat on a low setting until it is cool and creamy in texture. Pour the finished product into decorative containers and wait for it to cool completely before covering.

Store in a cool place (it may need to be stirred occasionally).

Shelf life is 6 to 8 months.

Gentle Facial Cleanser

Here is a cleanser gentle enough to use on your face every day. Be sure to shake it prior to use; then pour a small amount into your hands and rub them together to create some lather. Gently wash your face and rinse off with clear water.

½ cup (118 ml) liquid castile soap
½ cup (118 ml) distilled or purified water
2 tablespoons (30 ml), total, glycerine and rosewater blended together in equal amounts
2 tablespoons (30 ml) aloe vera gel

Combine all the ingredients and heat the mixture on the stove or in a microwave oven on medium heat until they have blended well. Allow the mixture to cool; then store in a decorative bottle. Shake prior to use.

Shelf life is 4 to 5 months.

Firming & Clarifying Facial Elixir

This wonderful elixir tones and firms without drying the skin; in fact, it actually moisturizes as it tones. Using a cotton ball, pat it onto your face and allow it to sit on the skin for about 10 to 15 minutes before rinsing off.

2 tablespoons (30 ml) glycerine
1 tablespoon (15 ml) witch hazel
1 tablespoon (15 ml) rosewater
2 tablespoons (30 ml) honey
1 tablespoon (15 ml) sweet almond oil
1 tablespoon (15 ml) wheat germ oil

Combine all the ingredients in a bowl, whisking or stirring them until they're smooth. Place the mixture in a pretty covered bowl, jar, or bottle and store it at room temperature. Shake prior to use.

Shelf life is 1 to 2 months.

Hint: *In place of rosewater, you can substitute an equal amount of any floral or herbal water. Just be sure you use ingredients that are safe for the skin.*

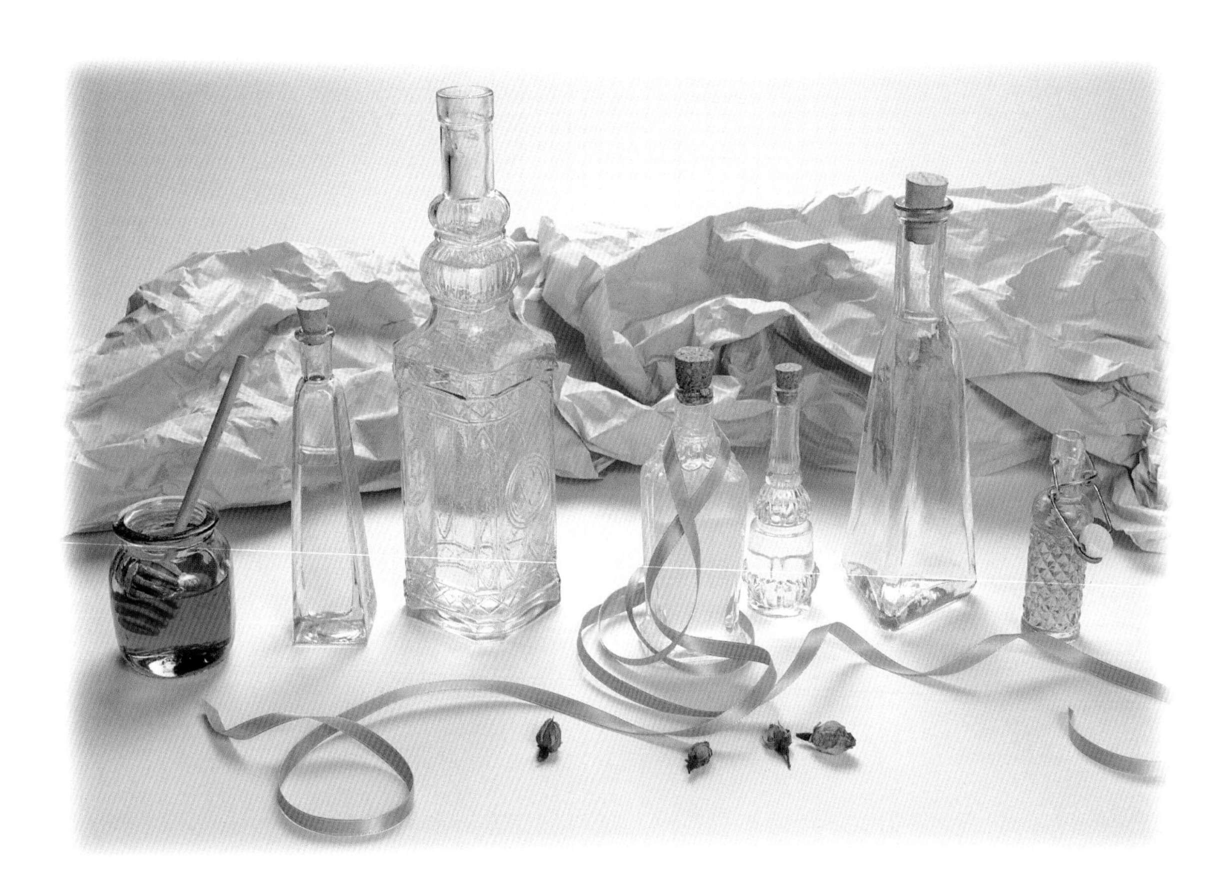

Vitamin E Under-Eye Moisturizer & Lip Balm

Are you a sun worshiper or have a friend who just can't resist the tropical look? If so, then you're certain to enjoy this moisturizing balm, which is formulated to soften those lines around the eyes. It can be poured into a small jar, or you can recycle a lipstick tube by cleaning it thoroughly and filling it with this "eye-saving" product. This recipe also makes an excellent lip balm.

1 teaspoon (5 ml) beeswax
1 teaspoon (5 ml) cocoa butter
1 tablespoon (15 ml) sweet almond oil
3 drops vitamin E oil
1 teaspoon (5 ml) paraffin wax*
1 teaspoon (5 ml) petroleum jelly*

In separate containers, melt each ingredient that needs to be melted; then combine all the ingredients in a double boiler, heating and stirring them until they are well blended. Pour the mixture into a lipstick pot, small jar, or old lipstick tube that you have lightly coated with almond oil. Let it sit for about 1½ hours; then refrigerate.

If the balm doesn't set up well enough, reheat it and add more wax and beeswax. If it's too dry, reheat and add more sweet almond oil.

Shelf life is at least 1 year.

Hint: *If you are making this as a lip balm not an under-eye treatment, add 1 drop of candy flavoring to make it even more enticing.*

*Although I don't usually recommend using these products, I made an exception in this case because of the need for extra protection on the lips or under the eyes.

Homemade Natural After-Shaves

After shaving, most people have many microscopic nicks (and sometimes a few not so microscopic) in the skin that may or may not be noticeable to the naked eye. These leave the skin open to infection. The purpose of an after-shave is to brace the skin—i.e., close the pores and protect the tiny cuts from bacteria and dirt.

Old-Fashioned After-Shave Bracer

This is a traditional after-shave, and it's included here just for fun and for your information. I prefer the next recipe because it replaces alcohol with gentler ingredients, but you may enjoy reading a more traditional formula first. Most commercial products consist simply of alcohol, water, preservatives, and a fragrance. (Ouch! You have to be very "manly" to stand them.) If you're making after-shave for a real "die-hard," this formula at least softens the effect of the alcohol. Otherwise, try the next recipe for a gentler product to pamper skin that may already be irritated by shaving.

¼ cup (59 ml) 90-proof vodka or isopropyl alcohol
¼ cup (59 ml) distilled water
¼ cup (59 ml) glycerine
¼ cup (59 ml) witch hazel
½ teaspoon (2.5 ml) essential oil (if desired)

Combine all the ingredients and place the mixture in attractive containers. Shake before using.

Shelf life is at least 1 year

Nondrying After-Shave

Unfortunately many people—men included—have sensitive skin and will often break out in a rash when alcohol is applied. (And this is in addition to the sting.) Here is a much gentler after-shave with glycerine to soothe the skin while still getting the job done. If you wish, you can add a drop or two of essential oil of lemon, lime, or cedar for a more pleasant fragrance, but limit this addition to no more than 3 or 4 drops (so we don't put back in the sting).

½ cup (118 ml) witch hazel
2 drops tea tree extract
½ teaspoon (2.5 ml) lemon juice concentrate
¼ cup (59 ml) distilled water
3 tablespoons (44 ml) glycerine

Combine all the ingredients and place the mixture in attractive containers. Shake before using.

Shelf life is 6 months.

Hint: *Keep either formula in the refrigerator for an extra-refreshing effect.*

Old-Fashioned Facial Toners & Astringents

Toners and astringents are traditionally used to remove excess cold cream or cleanser from the face. Commercial products are usually alcohol based (some even contain acetone—the chemical that removes nail polish) and are very drying to the skin. Because they contain more water, toners are gentler than astringents and are better for dry skin. Astringents are stronger so that they can remove more bacteria. They are often used for dealing with acne—the "more is better" approach—but over-drying the skin often leads to increased oil production, creating a vicious circle of events for teenagers. Here are some gentler, natural alternatives.

Peppermint Herbal Toner for Normal Skin

½ cup (118 ml) infusion of peppermint
½ cup (118 ml) witch hazel
½ cup (118 ml) water

Make the peppermint infusion by steeping 2 heaping tablespoons (about 32 ml) of peppermint leaves in 1 cup (237 ml) of hot water. Cover and let it sit for 15 minutes; then strain the liquid and measure out ½ cup (118 ml). Mix the measured portion of infusion with the other ingredients and pour the blend into a bottle. Add a dried sprig of mint for effect, if desired. Shake occasionally between uses.

Shelf life is 8 to 9 months.

Rose & Aloe Vera Toner for Dry Skin

½ cup (118 ml) witch hazel
½ cup (118 ml) water
4 tablespoons (59 ml) aloe vera gel
4 tablespoons (59 ml) rosewater
4 tablespoons (59 ml) glycerine

Mix all the ingredients together in a bottle. Shake occasionally and before use.

Shelf life is 8 to 10 months.

Astringent/Toner for Oily or Problem Skin

1 cup (237 ml) witch hazel
½ teaspoon (2.5 ml) tea tree extract
½ cup (118 ml) chamomile infusion

Make the chamomile infusion the same way you did the peppermint infusion in the recipe at left. Mix the infusion with the other ingredients and put the blend in a pretty glass bottle.

Shelf life is 8 to 9 months.

Variations: *To make a honey mint vinegar toner, put 1 tablespoon (15 ml) of honey in ¼ cup (59 ml) of hot water and stir well; then add this to ¼ cup (59 ml) of apple cider vinegar. Use this in place of the witch hazel in the formula at left. For a citrus peppermint toner, replace the witch hazel with the juice from half a lemon mixed together with sufficient water to equal ½ cup (118 ml).*

Facial Scrubs

The primary benefit of scrubs is their ability to exfoliate the skin, which removes dead cells and surface debris and encourages healthy new skin growth. Exfoliation promotes fresh, glowing skin and decreases the depth of wrinkles. This is a wonderful procedure, but you must use some caution. Don't get carried away and exfoliate too much, or you will get raw, irritated skin.

Honey Oatmeal Creamy Scrub

¼ cup (59 ml) uncooked oatmeal (not instant)

2½ teaspoons (12.5 ml) honey

1 teaspoon (5 ml) apple cider vinegar

½ teaspoon (2.5 ml) warm water

Combine all the ingredients and apply the mixture to your face as a mask, allowing it to sit for 10 minutes. Using your fingers or a soft cloth, apply some warm water and gently scrub the skin in a circular motion as you rinse off.

This scrub has no real shelf life; it must be used as it is made.

Almond Citrus Scrub

Unlike most masks and scrubs, this one has a long shelf life, making it perfect for gifts. It is suitable for oily and normal skin.

½ cup (118 ml) finely grated and dried citrus peels (totally dehydrated lemon, orange, or grapefruit peels)

½ cup (118 ml) blanched almonds

1 drop liquid face cleanser (see recipe on page 95 or use your favorite commercial cleanser)

Combine the citrus peels and almonds in a food processor or blender and process until the mixture is a fine powder. Store it in an attractive container and use it as follows: Place a small amount—about the size of a quarter—of the scrub in your hand and add one drop of cleanser and enough water to make a paste. Using a gentle circular and upward motion, rub the mixture onto your face for about 30 seconds in each area. Rinse with warm water. Follow with a toner and moisturizer.

Shelf life is at least 2 years if stored in a moisture-free environment.

Facial Masks

To make your own facial masks, you will need some French clay, fuller's earth, or bentonite clay, all available in powder form at your local health food store. Clay from the earth is a standard ingredient in most masks, and when you apply it to your skin together with water, it tightens your pores and temporarily firms your skin as it dries. As the clay hardens and tightens your pores, particles of dirt (blackheads) are forced out. This is a much safer process than squeezing them, as your fingernails can damage your skin. Various types of clay differ in the benefits they claim, according to their mineral content and area of origin, but all will tighten and firm the skin similarly. If you cannot find either fuller's earth or French clay, your pharmacist may be willing to order some for you.

Please don't attempt to make your own "earth clay" by digging it from your back yard. Even if your soil contains a lot of clay, this type of clay is not the same. Commercial clay has been cleaned and purified for cosmetic uses.

Note: *There is no real shelf life for any of these natural facial masks—you need to use them as you make them.*

Honey, Vinegar & Almond Mask

Use this mask for facial toning and softening. It's great for very dry skin because the honey and oil soften and moisturize, while the vinegar tones and helps adjust the pH.

2 tablespoons (30 ml) honey
¼ teaspoon (1.25 ml) vinegar
½ teaspoon (2.5 ml) sweet almond oil

Put the honey in a glass cup and microwave it just long enough to make the honey soft and pliable but not hot (20 to 30 seconds is usually more than enough). Stir in the other ingredients until you have a smooth blend and apply it immediately to your face. Leave it on your skin for 15 minutes before removing it with tissues and rinsing with water.

Yogurt Wonder Mud Mask

This mask is wonderful for oily or problem skin. If you want to give it as a gift, you can fill a small, attractive jar or sack with the powdered clay and accompany it with nicely printed instructions.

4 tablespoons (59 ml) fuller's earth or French clay
2 tablespoons (30 ml) water
2 tablespoons (30 ml) plain yogurt

Mix together the powdered clay, water, and yogurt to make a smooth paste. Apply the paste to your face immediately, using gentle, upward strokes and avoiding the area immediately around the eyes. Let the mask sit on your skin for approximately 15 minutes or until it's dry; then rinse it off with warm water.

***Hints:** If you don't have any yogurt on hand, mix 4 tablespoons (59 ml) of powdered clay with 5 tablespoons (74 ml) of water for a straight clay mask. If you want to use honey, mix 4 tablespoons (59 ml) of powdered clay together with 1 tablespoon (15 ml) of honey dissolved into 6 tablespoons (89 ml) of warm water; this mask is perfect for normal skin.*

Brewer's Yeast & Barley Water Mask

This mask is perfect for very oily or troubled skin. The powdered brewer's yeast can be found at health food stores and some pet supply stores (it helps ward off fleas). Most health food stores carry rolled or whole barley in bulk storage bins, allowing you to purchase it in small quantities.

¼ cup (59 ml) whole or rolled barley
¼ cup (59 ml) water
¼ cup (59 ml) powdered brewer's yeast

Put the barley and water in a pot on the stove and warm it to about 200°F (93°C). Remove the pot from the heat, cover, and let it stand overnight. In the morning, strain the water into a bowl and set the barley aside. Mix ¼ cup (59 ml) of brewer's yeast with 2 tablespoons (30 ml) of barley water and mix into a paste. Apply the paste to your skin and let it dry for 10 to 15 minutes. Rinse off and see how tightened and improved your skin feels.

Want to know how to use the leftover barley? Buy some unflavored yogurt, invite a friend to join you, and while you're doing your mask, she can make a barley yogurt mask for herself.

Barley Yogurt Mask

This mask is good for dry skin. It makes a perfect companion recipe for the brewer's yeast and barley water mask, which uses the barley-infused water you would otherwise throw away.

¼ cup (59 ml) barley
¼ cup (59 ml) water
¼ cup (59 ml) plain yogurt

Put the barley and water in a pot on the stove and warm it to about 200°F (93°C). Remove the pot from the heat, cover, and let it stand overnight. In the morning, strain off the water and combine the softened barley with the yogurt. Apply the mixture to your face and allow it to sit for 10 to 15 minutes; then wash it off. Your skin will feel soft and moisturized. If desired, apply a moisturizer to maintain your skin's softness.

Farmer's Kitchen Secret Mask

This formula is superb for all types of skin.

2 egg whites, chilled
⅛ teaspoon (0.6 ml) cornstarch

Whip the egg whites until they form peaks; then slowly add the cornstarch. Apply the mixture immediately to your face, leaving it on for 20 minutes, then rinsing it off. This is very effective for tightening the skin for a smoother face and a youthful glow.

Hawaiian Secret Avocado & Coconut Mask

This is another mask that is ideal for conditioning dry skin. Women in the tropics have long known the moisturizing properties of avocado and coconut. After applying the mask, try sipping some of your favorite fruit punch to get the full effect of this tropical experience.

1 medium avocado, mashed
Approximately 1 teaspoon (5 ml) coconut oil

Slowly add about 1 teaspoon or so (about 5 ml) of coconut oil to the mashed avocado and stir until it makes a creamy paste. Apply the paste directly to the skin and let it sit for 15 to 20 minutes before washing it off. Feel the soft, creamy texture it gives your skin!

Chapter Ten

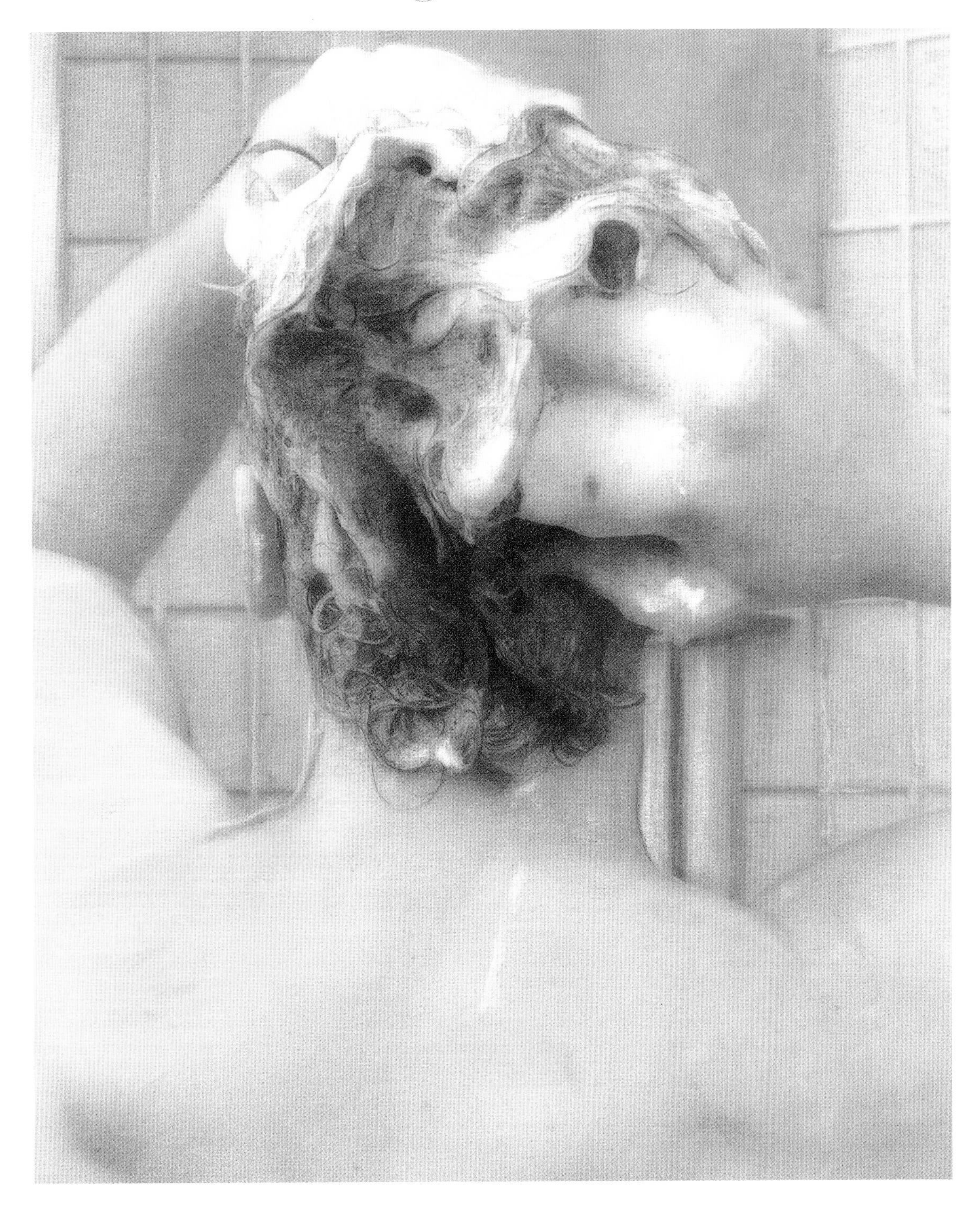

Hair Care Products

Ask any woman how she feels, and I can guarantee that part of her answer has at least a little (if not a lot) to do with what kind of a "hair day" she is having. For men and women alike, our hair is our crowning glory, and we gladly spend plenty of time and money caring for it. Here are some special hair treatments to whip up in your kitchen, and because each recipe describes the benefits of the major ingredients, you'll have plenty of room for exercising your own creativity.

Basic Glycerine Shampoo Formula

4 heaping tablespoons (about 60 ml) lye
1 cup (237 ml) cold soft water
⅔ cup (158 ml) glycerine, preferably vegetable glycerine, but animal-derived glycerine will also work
3 tablespoons plus 2 teaspoons (54 ml) coconut oil
2 cups (473 ml) lukewarm water
2 tablespoons (30 ml) commercial shampoo (optional)

While gently stirring, slowly add the lye 1 tablespoon (15 ml) at a time to the cold water. Read the warning on the label and be very careful when handling lye; it is very caustic and should not come in contact with your skin. *(Warning: This is not a procedure for children!)* It's best to wear rubber gloves and use a mask or avert your face while stirring to avoid inhaling any fumes, and be sure to stir slowly to prevent any spattering. This mixture will become hot! After you have stirred thoroughly, allow the mixture to sit for about a hour or until temperature drops to 90 to 95°F (32 to 35°C). Then heat the glycerine and coconut oil in a pan until the mixture reaches 150°F (65°C).

When both mixtures are at the proper temperature, slowly pour the glycerine and oil combination into the dissolved lye. Stir constantly until the combination reaches the consistency of honey, This will take about half an hour, depending on the outside temperature and other factors. While continuing to stir, slowly add the lukewarm water. This produces a very gentle, natural shampoo that cleans very effectively but has little lather. If you prefer more suds, add 2 tablespoons (30 ml) of your favorite commercial shampoo as a final step. Either mixture will set up fairly thick; it can be reheated to add more water, if desired, but it must be stirred well.

To vary the shampoo and give it fragrance, add a small amount of any of the following: aloe vera gel, jojoba oil, herbal infusions, honey, rosewater or other floral waters, natural extracts, aromatherapy oils, or dried herbs and spices. Experiment and have fun; just remember that a little fragrance goes a long way.

Shelf life is 4 to 6 months, with no special storage required.

Hint: *Hard water can be softened with washing soda or by using any commercial water softener (follow the manufacturer's instructions). If hard water is used, your shampoo may have a greasy film on top or may not mix evenly.*

Hint: *An alternative to making your own basic shampoo is to use commercial soap flakes dissolved in sufficient water to produce the desired consistency. (This procedure is safe for children to do.) Add fragrant or conditioning ingredients as before. If you don't find soap flakes on the shelf at your local pharmacy or health food store, they can be special-ordered or homemade by grating a bar of mild glycerine soap.*

Beer Shampoo

Beer helps give hair more volume and makes it appear thicker. Simply substitute ½ cup (118 ml) of beer for the same amount of water in the glycerine shampoo formula. Don't worry about your hair becoming overly dry; most of the alcohol is lost when the beer is heated during processing. If you're adding it to a commercial shampoo or the dissolved soap flakes, open the beer and let it sit for two days to become stale; this helps reduce the alcohol content naturally by evaporation.

Dandruff Treatments

Here are four natural approaches you can try for treating dandruff.

- Crush an aspirin tablet thoroughly and dissolve it in 2 teaspoons (10 ml) of warm water. Add this to your glycerine shampoo when you add the lukewarm water during the final step. Aspirin is a form of salicylic acid, an ingredient contained in many commercial dandruff shampoos, which helps dry the scalp and control dandruff.

- Add ½ teaspoon (2.5 ml) of tea tree extract during the final phase of making your shampoo and stir well. Tea tree is often used as a skin astringent or toner and is a natural ingredient harvested from the melaleuca tree, a native of Australia.

- Sweet birch oil from the Betula alba tree is harder to find than tea tree extract, but a friend from northern Europe insists that it is very effective. Sweet birch oil has astringent toning properties, and it's used in Europe for treating dandruff and eczema. Stir in ½ teaspoon (2.5 ml) during the final step of making your shampoo.

- Thyme, rosemary, and fir nettle extract are also good ingredients for helping dandruff. To make an herbal infusion for adding to your shampoo, boil ½ cup (118 ml) of water, add 1 tablespoon (15 ml) of herbs, cover, and let steep for 1 hour. Add 1 to 2 tablespoons (15 to 30 ml) to your shampoo, depending on the effect you desire. This alternative is not as potent as the tea tree or Betula alba methods.

Quick & Easy Egg Shampoo

This recipe is a combination of several of "grandmother's" formulas for beautiful hair. The egg adds protein to the hair to strengthen it, olive oil gives the hair gloss and sheen, and lemon juice works as a clarifier (allowing this formula to rinse clean). Water is included to dilute the strength of the first three ingredients to a good proportion for use on the hair, and the shampoo base provides the cleansing action. This is a great conditioning shampoo that you can whip up in your kitchen any time you want to give your hair a treat!

1 egg, lightly beaten
½ teaspoon (2.5 ml) olive oil
½ teaspoon (2.5 ml) lemon juice
½ cup (118 ml) water
1 cup (237 ml) glycerine shampoo, gentle commercial shampoo, or liquid castile soap

Using a whisk, mix all the ingredients together for 1 minute. Apply and massage gently into the hair. Rinse thoroughly.

This recipe can be kept in the refrigerator for about 1 to 2 weeks; just make sure to whisk the ingredients for about 1 minute prior to use.

Gentle Hair Coloring Aids

Plant ingredients can be used for tinting or preserving colored hair. Simply make an herbal infusion as described on page 113 for making an herbal infusion for treating dandruff. Add 2 to 3 tablespoons (30 to 44 ml) of the infusion to your shampoo.

Choose the herbs according to the hair color you wish to maintain

- Brunettes: Rosemary, sage, or black malva flowers
- Blonds: Chamomile or yarrow
- Redheads: Red sage or madder root

Henna, a natural ingredient derived from the bark of a tree, is very effective in coloring your hair red. It is also very strong, so use this product in moderation. Most health food stores and some drug stores carry coloring products containing henna.

Super Hair Conditioning Pack

This is an after-shampoo hair conditioner for all types of hair. The egg and gelatin provide strength and nourish the hair with protein, the fruit moisturizes, and the oils provide shine and gloss. A small amount of water is included to make a good consistency and to shorten the rinsing time required.

1 egg, slightly beaten
½ avocado or ½ banana, peeled and mashed
1 tablespoon (15 ml) wheat germ oil
1 tablespoon (15 ml) aloe vera gel
2 tablespoons (30 ml) apricot kernel oil
1 tablespoon (15 ml) unflavored gelatin
¼ cup (59 ml) water

Mix all the ingredients together; then use your fingers to work the mixture through your hair thoroughly. Cover your hair with a plastic shower cap or plastic wrap and leave it on for 15 to 20 minutes. Shampoo well with a gentle shampoo.

Shelf life is short—up to 1 week if kept in the refrigerator and stirred prior to use.

Hot Oil Treatment

This makes a good conditioner for dry scalp and hair, and all the ingredients add gloss and shine to the hair.

½ cup (118 ml) sweet almond oil
¼ cup (59 ml) apricot kernel oil
¼ cup (59 ml) olive oil

Mix all the ingredients together. Shampoo your hair and towel it dry. Warm the oil mixture in the microwave or on the stove top until it reaches 95°F (35°C). Then apply the mixture in small amounts and comb it through your hair, using your fingers to work it into the scalp. Cover your hair with a plastic shower cap or plastic wrap and leave it on for 15 to 20 minutes. Alternatively, you can apply the oil mixture at room temperature and sit under the hair dryer for 15 to 20 minutes to activate the treatment. Afterward, cleanse your hair well with a gentle shampoo.

Shelf life is approximately 1 year.

Old-Fashioned Hair Spray

The active ingredient in this formula is lemon juice, which causes the hair to stiffen slightly, giving it a little more "hold." It can also lighten your hair over time, so keep this in mind as you use it. This is an added benefit if you want your hair to have that "sun-kissed" look in the summertime, and the lemon juice is much gentler than bleach. However, lemon juice can dry your hair, so you may want to use a conditioning treatment as well. Remember that more is not always better—don't put straight lemon juice on your hair.

1 whole lemon
1 cup (237 ml) distilled or purified water

Slice the lemon, add it to the water in a pan, and boil. After it has cooled, strain the solution and put it in a pump bottle. Refrigerate between applications.

Shelf life is about 1 month.

Chapter Eleven

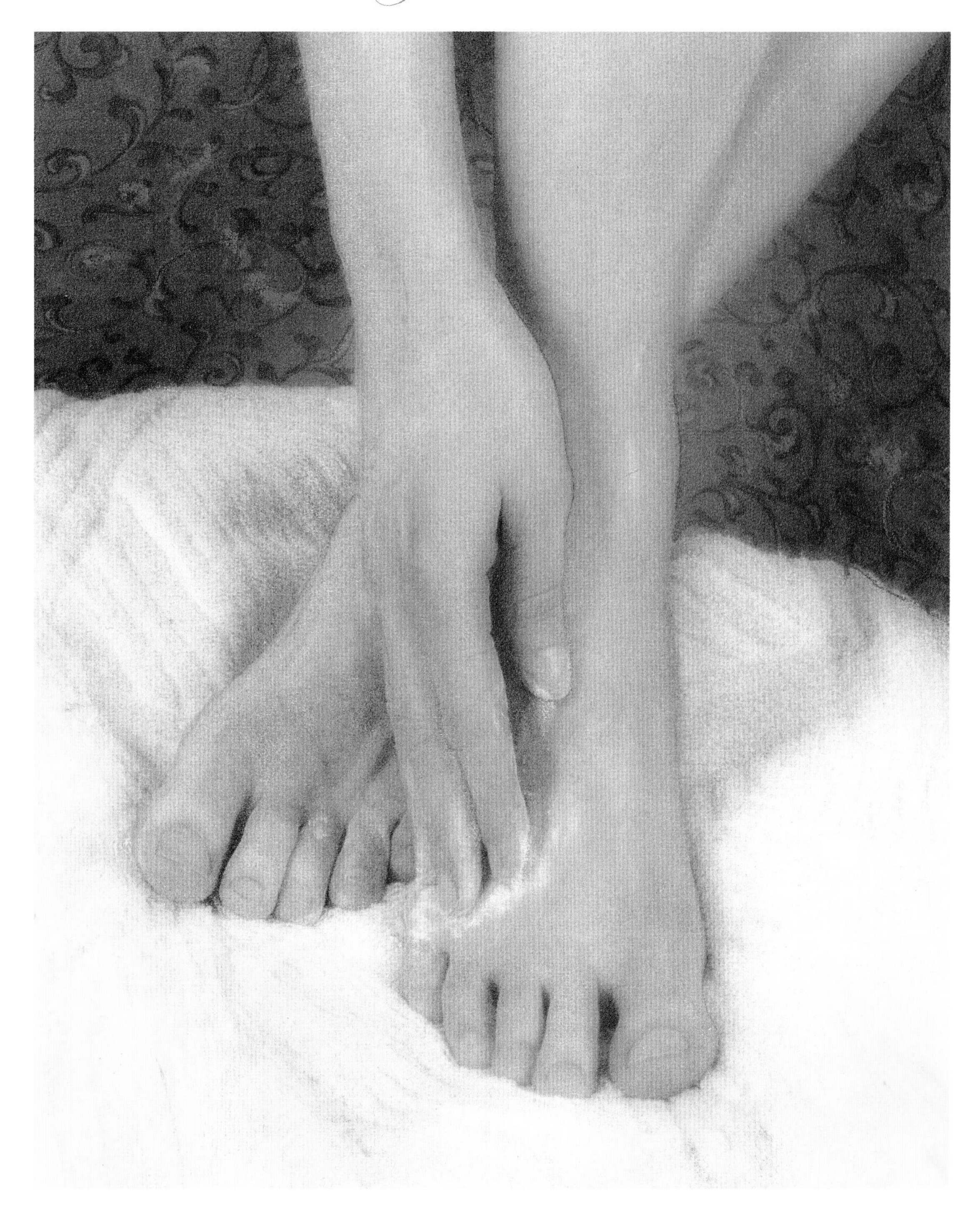

Especially for Hands & Feet

OUR HANDS ARE OUR MOST VALUABLE tools, and they're in and out of everything we do all day, subject to countless irritations from water, wind, and whatever else we touch. Likewise, we often forget to care for our poor, aching feet, and yet they serve us all day long. Both deserve to be pampered with these natural, soothing treatments.

Hand Care

Artists, dishwashers, gardeners, farmers, and mechanics—just to name a few—are all guilty of "hand abuse," but we all tend to think about our hands only after the damage is done. If you find yourself with dry, cracked, and sore hands, some homemade relief is just what you need.

Salon-Style Paraffin Hand & Cuticle Treatment

When you apply oil to your hands and follow it with a coating of melted wax mixed with oil, the warm wax presses the oil into the skin, softening it. A process similar to this is used in many beauty salons prior to a manicure.

½ lb (227 g) paraffin wax
4 tablespoons (59 ml) sweet almond oil

Using a low setting on your stove, heat the paraffin in a pan until it's melted. Put a small amount of the oil on your hands and rub it into the skin. Then pour the rest of the oil into the paraffin and stir. Remove the pan from the heat and keep stirring. When the wax is not yet set up but is cool enough to touch, dip your whole hands into it. Let the wax cool for 5 minutes; then peel it off to reveal smoother, softer, well-moisturized hands.

Save any unused wax mixture for another application later. You can pour it into an old ice tray and pop out the cubes when they have set up. (If the cubes seem too soft, try adding a little less oil to your next batch.) Wrap them with bows and give them as gifts with instructions for your friends and family. Their hands will thank you.

Shelf life is 1 year or more.

Variation: *If you prefer to use beeswax in place of the paraffin, decrease the amount of oil to about 3 tablespoons (44 ml).*

Deep, Intensive Hand Treatment

This treatment is like a sauna just for your hands. A combination of oils actively softens the hands overnight, while a layer of plastic wrap keeps the skin moist and a pair of gloves provides warmth for the oils to work more effectively.

4 tablespoons (59 ml) sweet almond oil
4 tablespoons (59 ml) coconut oil
4 tablespoons (59 ml) olive oil
plastic wrap
1 pair cotton gloves

Melt the coconut oil over low heat and combine the oils right before going to bed at night. Generously apply the mixture to your hands, wrap both hands in plastic wrap, and put on the gloves. Continue wearing the gloves overnight—you won't believe the improvement in the morning. If you repeat this process every night until you've used all of the oil mixture, you may not recognize your own hands.

Shelf life is about 3 months.

Hand Butter

This butter will be firm when it sets up, but it will soften and melt in your hands when it's used.

¼ cup (59 ml) cocoa butter
¼ cup (59 ml) beeswax
2 tablespoons (30 ml) safflower oil
4 to 6 drops of a fruity essential oil, if desired

Warm the cocoa butter and beeswax in a pan until they're melted. Add the safflower oil and remove the mixture from the heat. Add the essential oil now, if you wish. Using a whisk, beat the mixture until it's cool. While it's still a little soft, place the hand butter in a decorative container that offers easy access to your fingers.

Shelf life is 6 months or more.

Cuticle Treatment

4 tablespoons (59 ml) castor oil
4 tablespoons (59 ml) cocoa butter

On low heat, warm the cocoa butter until it is liquid. Add the castor oil and stir. Remove from the heat and put the mixture in a deep bowl. When the liquid is cool enough, soak your fingertips and nails for 10 to 15 minutes. Rinse your fingers in warm water and push back the cuticles. Put any excess into a pretty container to save for a future treatment.

Shelf life is approximately 3 months.

Caring for Your Feet

Wrapped all day long in leather or plastic, feet get squeezed, jarred, and stood upon for long hours. In the summer, they're exposed to burning sands, and in the winter they suffer cold, damp conditions. If you pamper them with some of these homemade treatments, you will be rewarded with beautiful, healthy feet.

Callous Remover

3 tablespoons (44 ml) table salt
3 tablespoons (44 ml) fine sand, ground pumice, or loofa bits (you can grate a loofa with a sharp cheese grater)
1 teaspoon (5 ml) liquid shower gel, shampoo, or soap
A few drops essential oil, if you wish

Using a spoon, mix all the ingredients together in a bowl. Dip a washcloth into the mixture and scrub your callouses to remove them. Do not overdo this procedure; callouses should be removed a little at a time.

This has a shelf life of about 6 months. You can store it in a decorative container, but you will need to stir it again before use.

Peppermint & Menthol Foot Salts

After a long day of being on your feet, there's nothing like a refreshing foot soak to make you feel like a new person!

¼ cup (59 ml) epsom salts
¼ cup (59 ml) sea salt
4 drops liquid menthol
4 drops peppermint essential oil

In a large sink or small bathtub (a large bucket works well too), place just enough very warm water to cover your feet. Add all the ingredients and swirl them into the water with your hand. Sit down in a comfortable position, put your feet in the water, and enjoy!

If you don't want to use these foot salts right away, place all the ingredients together in a plastic bag and shake them to distribute the essential oils. Then place them in a decorative container, where they will last up to a year.

Variations: *Substitute eucalyptus for the menthol or use rosemary in place of peppermint.*

Hint: *In hot weather, use very cool water to dissolve your salts and soak your feet.*

Eucalyptus Foot Deodorizer

To help keep your feet fresh all day, apply this deodorizer right after bathing in the morning.

½ cup (118 ml) cornstarch
½ cup (118 ml) baking soda
6 drops eucalyptus oil or peppermint essential oil

Using a hand sieve or sifter, combine the powders and add the essential oil a little at a time. Sift through many times to force the essential oil through the powder.

Shelf life is 1 year to 18 months.

Peppermint Foot Freshener

You can spray this refreshing formula on your feet for a quick pick-me-up any time of the day or evening.

½ cup (118 ml) witch hazel
12 drops peppermint essential oil

Combine the ingredients and put the mixture in a decorative spray bottle, labeling it with instructions to shake before use.

Variations: *If you want a more powerful cooling effect, eucalyptus oil can be substituted for the peppermint, and isopropyl alcohol can be used in place of the witch hazel. Be sure to follow with a moisturizer, as this will leave the feet feeling dry.*

Shelf life is 6 to 8 months, sometimes longer.

Chapter Twelve

A Dozen of Grandma's Natural Recipes

JUST FOR FUN, HERE ARE SOME OF MY favorite old-time solutions for a variety of life's little irritations. Some were found in my grandfather's library of early medicine and folklore; others came straight from my mother and grandmother and were handed down through my family. Believe it or not, these recipes really work—some for reasons I know, and others for reasons I have yet to learn. I hope you enjoy them as much as I do!

Eliminating Puffiness Under the Eyes

Fresh chilled cucumber slices (preferably with the skin peeled off) or moistened chamomile tea bags work equally well for this problem. Lie down, with your feet up, and apply either remedy directly to your closed eyes for 15 to 20 minutes. Not only does this reduce swollen tissue around the eyes, but it also feels wonderful. Cucumber is known to be a soothing and toning agent, and the coldness helps reduce the puffiness too. Why the chamomile works is less certain; my theory is that it functions as a poultice, much the way tobacco leaves were once used by growers to pull out infections.

Relieving Sunburn Pain

According to my mom, who *swears* this to be true, apple cigar vinegar eases the discomfort of too much exposure to the sun. Many a summer night I was sent to bed smelling like a salad, but I am a witness to the fact that this remedy—if applied in time—really works.

For apple cider vinegar to be effective, you must apply it directly to the skin as a new burn is developing. Use a tissue or cotton balls to dab it gently onto the skin. The burn will continue to intensify, but the pain won't. If you wait too long after the burn has developed on the skin, the remedy won't be effective.

If you miss your opportunity with the apple cider vinegar and the burn has fully developed for 12 hours or more, then apply aloe vera concentrate in the same way (dab it onto the skin), and this will help relieve your discomfort.

Soothing Cooking-Related Burns

Most of us know that we should immediately run cold water on a burn. This is a good idea because the skin can continue to be affected by the heat of the burn itself, which increases the injury. Cold water helps cool the area quickly and limit the damage.

What many don't know is that a fresh slice of raw potato can aid the healing process. When I was young and just beginning to cook, my hand once got splattered with some hot grease. My mother, who happened to be present, quickly grabbed my hand, put it under cold running water for about a minute, then sliced a fresh potato in half and put the open end of the potato on my burn for five minutes. The burn never did hurt during healing! My research shows that the starch from the potato was the agent responsible for soothing the burn. (Keep in mind that this works only for minor injuries; serious burns need medical attention.)

Insect Repellent Lotion

Do you find yourself spending lots of money on bug repellent to ward off mosquitoes? Instead, make your favorite lotion or cream (or use an unscented commercial brand) and add 4 to 5 drops of citronella essential oil per ½ cup (118 ml) of lotion. To make the lotion even more effective for a wider variety of insects, add 3 drops each of thyme essential oil and pennyroyal essential oil. This concoction won't smell too appealing, but it is effective.

Caution: Pennyroyal can be toxic to pets and children if they ingest too much of it. If this is a concern, omit the pennyroyal.

Freckle & Age Spot Reducer

To your favorite lotion or cream, stir in several drops of pure lemon juice. (Too much juice will thin your product, so you will need to experiment with the exact amount.) This creates a

natural skin bleach. For a stronger solution, you can make a 50-50 blend of lemon juice and water and blot it directly onto your skin with cotton balls. You will find that lemon juice really tightens your pores as well as lightens your skin. I prefer to apply it in a lotion or cream because lemon also dries your skin. If you choose to use the 50-50 blend, make sure that you moisturize afterward.

Dry Throat & Cough Syrup

Combine the juice of a whole lemon, 3 tablespoons (44 ml) of glycerine, ½ cup (118 ml) of honey, and ¼ cup (59 ml) of water. Heat sufficiently to mix the ingredients together, stir, and bottle. The shelf life of this syrup is 1 to 2 months. If it becomes too cold to pour easily, it can be heated again before using. The lemon provides vitamin C (to help fight any viruses); the honey and glycerine soothe, coat, and protect the throat; and the water provides the correct dilution for the lemon juice. This home formula really helps with minor throat irritations and dry coughs.

Breath Freshener

When used as a mouthwash, peppermint tea (an infusion made with dry mint leaves) makes a gentle sweetener for your breath. For a stronger version, add 6 drops of peppermint extract to ¼ cup (59 ml) of water. Dip your toothbrush into the minty solution and brush your teeth, tongue, and insides of your cheeks with it.

If your predicament stems from sores or gum problems, another old-fashioned remedy is to brush your teeth with a 50-50 blend of baking soda and salt. This recipe was used in grandma's day for gum problems and as an oral toner.

Removing Lemon, Onion, or Garlic Smell from Hands

This recipe is especially appropriate since this book contains several others using lemon. It also helps eliminate onion and garlic smells! The secret ingredient is a handful of coffee beans.

Make or buy a small, loose-weave fabric or gauze pouch and fill it with fresh coffee beans. After working with lemon, onion, or garlic, wash and dry your hands; then rub them all over with the coffee-filled pouch. The offensive smell will go away much more quickly than with washing alone. You will need to replace your beans every 6 months or so to keep this remedy fresh and effective.

Alleviating Bee Stings & Bug Bites

Mix salt and baking soda in equal amounts. Then add enough water to make a paste. When dabbed onto a sting or bite, this will help lessen the redness and itching.

Moisturizing Nervous, Dry Mouth

Do you have an occasion where you have to give a speech or need to smile all day? Most people get dry-mouthed when they become nervous, and their lips stick to their teeth, preventing them from smiling and talking naturally. If this happens to you, apply a light coating of petroleum jelly over your teeth, and it won't matter how dry your mouth is or how nervous you feel. That smile will glide into place with confidence.

Taking the Red Out of a Blemish

Combine 1 teaspoon (5 ml) of lemon juice, 1 teaspoon (5 ml) of apple cider vinegar, and 1 teaspoon (5 ml) of salt. Apply the solution directly to the reddened area and leave it on for 15 minutes. Rinse. This helps draw out the inflammation from the infected area and gently bleaches it at the same time.

Keeping Cool in Hot Weather

Use the recipe from this book to make your own dusting powder and follow an old southern tradition—also practiced in Japan—for keeping cool on hot nights: sprinkle some powder on your sheets before retiring to help absorb excess perspiration while you sleep. (This works equally well for women suffering hot flashes at night.)

Do your feet perspire? Sprinkle some of the powder in your shoes first thing in the morning.

If you need additional help for your deodorant or just want to freshen up before going to bed, then dust some powder under your arms.

Applying powder has become a lost tradition because many people are concerned about the potential presence of asbestos in talc. If you use your homemade powder with its gentle alternative ingredients, this need not be a factor.

Chapter Thirteen

Packaging & Presentation

NO MATTER HOW MUCH CARE and effort you put into your homemade beauty products, they will not have the desired impact unless you package them effectively. A beautiful glass bottle adds mystique to a relatively colorless massage oil, and an old-fashioned tin makes a lip gloss all the more inviting. If you plan to give an assortment of products as a gift, arrange them attractively in a wire basket, wooden box, glass bowl, or decorative tin to make them look really professional.

Packaging Ideas

Fancy tins, specialty jars, small crocks, plastic tubs, and beautiful glass bottles all make good individual containers for your skin care products and other beauty formulas. Garage sales and estate sales provide some of the best opportunities to find one-of-a-kind containers. For larger quantities of specific shapes and sizes, kitchen supply stores and the kitchen sections of department stores often have good selections. Some floral supply stores also carry interesting containers. If you plan to make up quite a bit of one or more products, a commercial bottle company may be willing to sell you containers by the case.

Cleanliness is a must for all the containers you use. If you want to recycle an antique bottle or old-fashioned jar, make sure you wash it thoroughly and remove any old labels. Some items can be adequately sanitized in your dishwasher; other containers may require you to use a baby bottle brush or other specialized device to scrub them thoroughly.

Never take a chance if you're not sure whether a container is clean—discard the jar or bottle or use it for something other than your natural beauty products.

Remember, cleanliness is just as important for your skin care products as it is with food items. Once you have cleaned them thoroughly, your containers can be stored for use later. If you want to be extra cautious, clean them once again just before using them to hold your creams, lotions, and other beauty formulas.

You may also want to include small self-adhesive labels on your products. Rectangular or oval-shaped decorative labels can be found at stationery stores and in the greeting card areas of some department stores. Using a pen with permanent ink, write the name of the product and its ingredients on each label. For the convenience of the user, indicate the date packaged and the shelf life of the product.

Gift Presentations

Gifts should look distinctive, even if they're being given for no special reason. With just a modest investment in boxes, baskets, and other containers, you can make your gift assortments truly memorable. If you want to give a collection of beauty products to a friend or relative on a special occasion, choose the contents and the packaging materials carefully so that they complement each other. Here are a few combinations you might try.

MOTHER'S DAY: For that special mom, include a moisturizing lotion or cream, bath crystals (in a pretty jar or tied into a fabric sachet), lip balm, facial mask, and a large natural sponge. Arrange these in a crystal or silver bowl or in a beautiful basket with lots of ribbon and lace. A small cluster of flowers—fresh or dried—adds a nice touch.

SECRETARY'S DAY: To ease those tired feet at the end of a long work day, what could be more welcome than a collection of foot care items? Foot rub, foot powder, stress-relieving bath salts, foot scrub, and a loofa all make a perfect combination for any working person who wants to pamper his or her feet at the end of a long day. A wire basket makes a tongue-in-cheek container (an "in-box"); use colored tissue paper, shredded office paper, or colorful cellophane as packing materials.

FATHER'S DAY: Some natural veggie soap, nondrying after-shave, a lightly scented massage cream, massage oil, and egg shampoo are ideal for any man in your life. If he shaves with a blade, not an electric razor, include a natural bristle shaving brush too.

Present these in a container that reflects his favorite interests—a small tool box, garden trug, or golfing hat.

HOUSEWARMING: To soothe the shattered nerves of someone who has just gone through the ordeal of moving, this is a very welcome collection: stress-relieving bath salts, comforting aromatherapy essential oils, a floral bath vinegar, lip balm, and a few decorator soaps. Present these in a wicker basket or decorative bowl, lining it with a finely woven cotton or linen hand towel.

VALENTINE'S DAY OR BRIDAL SHOWER GIFT: For your valentine sweetheart or friend

about to be married, make up a package that includes heart-shaped soaps, rosewater and glycerine skin softener, massage oil, moisturizing bubble bath, and vanilla essential oil. Include a few candles (red for Valentine's Day and white for the prospective bride) and package this in an ice bucket with a bottle of champagne and two champagne glasses.

BABY SHOWER: Extra-fine body powder, body oil, gentle chamomile soap (use either the veggie or the glycerine soap recipe and add dried chamomile), and cucumber and aloe cream all make safe and gentle gifts for a baby. Place them in a pink, blue, and white hatbox or in a colorful plastic sand pail that has been lined with a baby towel that drapes attractively over the sides.

CHRISTMAS: A basket made of red and green fabric can be filled with holiday-scented soaps, citrus and spice essential oils (such as tangerine, cinnamon, and clove), pinecones scented with essential oils, chamomile (soothing) or citrus (toning) bath vinegar, and cinnamon massage oil. This makes a gift that not only adds holiday aromas to the house but also helps the recipient cope with the extra stress of the season.

HOSPITAL VISIT: Soothing cucumber and aloe cream, extra-fine body powder, glycerine shampoo, homemade soap, and moisturizing lip balm are all welcome additions to cheer up a hospital setting. Package them in a little wooden crate or box that can also hold the patient's watch or other personal items near the bedside.

Subject Index

Recipe Index